**Oklahoma**

**HOLT Social Studies**

# World Geography
## Test Preparation Workbook

**Preparation for the OCCT**

**HOLT, RINEHART AND WINSTON**
A Harcourt Education Company
Orlando • Austin • New York • San Diego • London

Printed in the United States of America



ISBN 0-03-092298-4

4 5 6 7 179 09 08 07

# Contents

# Test-Taking Tips for Students

When you take a standardized test in Social Studies, you will answer multiple-choice questions about Social Studies topics. Use these suggestions to help you do well on standardized tests such as the OCCT.

### 1. Study the Directions

Read all the directions carefully. Then, study the answer sheet. Be sure you know exactly what to do before you make the first mark.

### 2. Read Questions Carefully

Read every question carefully. Look for negative words, such as *never, unless, not,* and *except.* If a question contains one of these words, look for the answer that does not fit with the other answers.

### 3. Answer Every Question

You will not be penalized for guessing. If you are not sure of an answer, circle the question and move on. Go back to the question later. If you are still uncertain of the answer, use the 50/50 strategy and make an educated guess. Read every choice carefully. Then, eliminate the two answers you think are least likely to be correct. If two choices seem equally correct, make an educated guess.

### 4. Get the Full Picture

When a question refers to a map, chart, or graph, read all the information carefully, including headings and labels. Determine any trends or oddities before answering the question.

### 5. Check Your Work

You are not finished with your test until you check it. Make sure you haven't left any answers blank or put two answers to one question. Review the hardest questions and erase any smudges or stray pencil marks.

Name ______________________ Class ______________ Date ______________

## A Geographer's World

## Chapter 1 Test

For each of the following, mark the letter of the best choice.

**1.** Geographers look at the world
- ○ **A** by studying cities first.
- ○ **B** at the local, regional, and global levels.
- ○ **C** by studying only its physical features.
- ○ **D** as separate regions with no effect on each other.

**2.** A geographer's tools include
- ○ **A** maps and globes.
- ○ **B** satellite images.
- ○ **C** notebooks and tape recorders.
- ○ **D** all of the above.

**3.** Which of the following is an essential element of geography?
- ○ **A** uses of geography
- ○ **B** latitude
- ○ **C** movement
- ○ **D** relative location

**4.** The two main branches of geography are
- ○ **A** regional and local.
- ○ **B** cartography and meteorology.
- ○ **C** the study of water and the study of landforms.
- ○ **D** physical geography and human geography.

**5.** A social science is a field that studies people and
- ○ **A** geography.
- ○ **B** their relationships.
- ○ **C** their environment.
- ○ **D** the landscapes they create.

**6.** Which of the following questions might a geographer studying at a local level ask?
- ○ **A** Which products does a country export?
- ○ **B** What languages do people speak in South America?
- ○ **C** How do people in a town or community live?
- ○ **D** What features make up a physical region?

**7.** Globes can show

- ○ **A** specific information about a place, such as languages spoken or street names.
- ○ **B** a flat representation of Earth's surface.
- ○ **C** the world as it really is.
- ○ **D** more information than maps.

**8.** Which theme of geography describes features that make a site unique?

- ○ **A** human-environment interaction
- ○ **B** movement
- ○ **C** place
- ○ **D** location

**9.** Your street address describes

- ○ **A** absolute location.
- ○ **B** relative location.
- ○ **C** your environment.
- ○ **D** your society.

**10.** Which of the following would a human geographer study?

- ○ **A** soils and plants
- ○ **B** maps and globes
- ○ **C** weather and climate
- ○ **D** economics and politics

**11.** Geography is considered a social science because it

- ○ **A** deals with people and how they live.
- ○ **B** can be studied using numbers.
- ○ **C** uses satellite imaging.
- ○ **D** is studied in all countries of the world.

**12.** Two basic types of satellite images are

- ○ **A** full color and infrared.
- ○ **B** true color and infrared.
- ○ **C** full color and heat.
- ○ **D** heat and true color.

Name ______________________ Class ______________ Date ______________

**Planet Earth**

# Chapter 2 Test

For each of the following, mark the letter of the best choice.

**1.** Which theory suggests that the continents were once part of one supercontinent?
- ○ **A** tilt and rotation
- ○ **B** continental drift
- ○ **C** Ring of Fire
- ○ **D** eruption patterns

**2.** The main processes of the water cycle are
- ○ **A** evaporation and precipitation.
- ○ **B** drought and flooding.
- ○ **C** salt water and freshwater.
- ○ **D** groundwater and surface water.

**3.** Which is *not* a way that people change landforms?
- ○ **A** drilling tunnels through mountains
- ○ **B** building dams
- ○ **C** building terraces for farming
- ○ **D** creating sediment

**4.** Which of the following affects the amount of solar energy the planet receives?
- ○ **A** precipitation
- ○ **B** revolution
- ○ **C** plate collision
- ○ **D** plate separation

**5.** What regions have seasons marked by rainfall rather than temperature?
- ○ **A** high latitudes
- ○ **B** mid-latitudes
- ○ **C** areas near the South pole
- ○ **D** the tropics

**6.** During which season does solar energy begin to increase?
- ○ **A** spring
- ○ **B** summer
- ○ **C** fall
- ○ **D** winter

**7.** Monsoons are
- ○ **A** seasonal winds that bring heavy rains.
- ○ **B** volcanic eruptions.
- ○ **C** weathered landforms.
- ○ **D** low latitudes.

**8.** What parts of Earth can receive up to 24 hours of sunlight a day?
- ○ **A** areas near the equator
- ○ **B** areas with low latitudes
- ○ **C** areas near the Arctic and Antarctic circles
- ○ **D** areas tilted away from the sun

**Study the map below and answer the question that follows.**

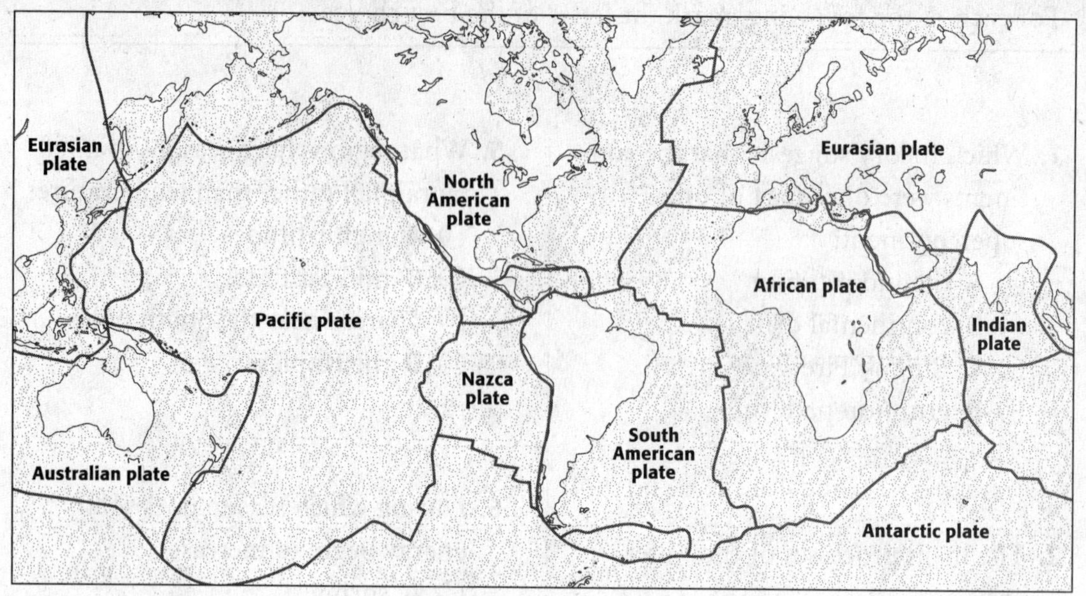


**9.** Two continental plates that seem to have collided are the
- ○ **A** North American plate and South American plate.
- ○ **B** African plate and Australian plate.
- ○ **C** Eurasian plate and Indian plate.
- ○ **D** South American plate and Pacific plate.

**10.** Every 24 hours, Earth completes one
- ○ **A** revolution.
- ○ **B** axis.
- ○ **C** tilt.
- ○ **D** rotation.

**11.** Water shortages can result from
- ○ **A** condensation.
- ○ **B** precipitation.
- ○ **C** drought.
- ○ **D** runoff.

**12.** Which of the following creates Earth's change in seasons?
- ○ **A** tilt
- ○ **B** rotation
- ○ **C** solar energy
- ○ **D** latitude

Name ______________________ Class ______________ Date ______________

## Climate, Environment, and Rescue

## Chapter 3 Test

For each of the following, mark the letter of the best choice.

**1.** Which is the best description of the Mediterranean climate?

- ○ **A** cloudy, mild summers and cool, rainy winters
- ○ **B** sunny, warm summers and mild, wet winters
- ○ **C** humid, hot summers and cold, dry winters
- ○ **D** humid, hot summers and cool, rainy winters

**2.** Which of the following is a renewable resource?

- ○ **A** natural gas
- ○ **B** gemstones
- ○ **C** forests
- ○ **D** petroleum

**3.** Which type of climate can occur at many different latitudes?

- ○ **A** tundra
- ○ **B** highland
- ○ **C** tropical savanna
- ○ **D** subarctic

**4.** At about 30° North and South latitude we find most of the world's

- ○ **A** tornadoes.
- ○ **B** rain forests.
- ○ **C** deserts.
- ○ **D** ecosystems.

**5.** One disadvantage of nuclear power is that it

- ○ **A** is a renewable resource.
- ○ **B** cannot be used to heat homes.
- ○ **C** pollutes the air.
- ○ **D** produces dangerous waste.

**6.** What happens when an environment changes?

- ○ **A** Habitats are often destroyed.
- ○ **B** Prevailing winds shift to the opposite direction.
- ○ **C** Nonrenewable resources are used even faster.
- ○ **D** Humans fix the problem by introducing new species.

**7.** The spread of desertlike conditions is called
- ○ **A** desertification.
- ○ **B** steppe.
- ○ **C** deforestation.
- ○ **D** permafrost.

**8.** Which of the following can be made into cosmetics and plastics?
- ○ **A** quartz
- ○ **B** bauxite
- ○ **C** petroleum
- ○ **D** humus

**9.** Highland climates change
- ○ **A** with the seasons.
- ○ **B** with elevation.
- ○ **C** according to precipitation.
- ○ **D** once every ten years.

**10.** The amount of sun at a given location is affected by Earth's
- ○ **A** tilt.
- ○ **B** movement.
- ○ **C** shape.
- ○ **D** all of the above

**11.** Ecosystems
- ○ **A** are any materials found in nature that people use or value.
- ○ **B** are made up of renewable and nonrenewable resources.
- ○ **C** can be of any size and can occur wherever air, water, and soil support life.
- ○ **D** move heat around Earth through large streams of surface water.

**12.** The two dry climate types are
- ○ **A** desert and steppe.
- ○ **B** desert and highland.
- ○ **C** steppe and highland.
- ○ **D** Mediterranean and desert.

Name ______________________ Class ______________ Date ____________

## The World's People

# Chapter 4 Test

For each of the following, mark the letter of the best choice.

**1.** A group of people who share a common culture and ancestry are a/an
- ○ **A** culture region.
- ○ **B** ethnic group.
- ○ **C** democracy.
- ○ **D** culture trait.

**2.** Which economic system is based on free trade and competition?
- ○ **A** command economy
- ○ **B** communism
- ○ **C** developed countries
- ○ **D** market economy

**3.** Assistance provided to people in distress is called
- ○ **A** interdependence.
- ○ **B** migration.
- ○ **C** humanitarian aid.
- ○ **D** cultural diversity.

**4.** Which is the economic activity that uses raw materials to produce or manufacture something new?
- ○ **A** primary industry
- ○ **B** secondary industry
- ○ **C** tertiary industry
- ○ **D** quaternary industry

**5.** What is the process of moving from one place to live in another?
- ○ **A** globalization
- ○ **B** cultural diffusion
- ○ **C** migration
- ○ **D** innovation

**6.** The term for having a variety of cultures in the same area is
- ○ **A** cultural diversity.
- ○ **B** culture region.
- ○ **C** popular culture.
- ○ **D** population density.

**7.** An area in which people have many shared culture traits is called
- ○ **A** a culture region.
- ○ **B** cultural diffusion.
- ○ **C** an ethnic group.
- ○ **D** popular culture.

**8.** Which type of government owns all property and dominates all aspects of life in a country?

- ○ **A** democracy
- ○ **B** monarchy
- ○ **C** dictatorship
- ○ **D** communism

**9.** What is the total number of people in a given area?

- ○ **A** birthrate
- ○ **B** culture
- ○ **C** population
- ○ **D** natural increase

**10.** People's culture

- ○ **A** never changes.
- ○ **B** is handed down by families and affected by history and new ideas.
- ○ **C** is the same all around the world.
- ○ **D** is left behind when people migrate.

**11.** A country's population can grow through

- ○ **A** a low birthrate and high death rate.
- ○ **B** cultural diffusion.
- ○ **C** natural increase and migration.
- ○ **D** globalization.

**12.** Which of the following is ruled by a single person with all the power?

- ○ **A** communist government
- ○ **B** democracy
- ○ **C** dictatorship
- ○ **D** command economy

Name ______________________ Class ______________ Date ______________

## The United States

## Chapter 5 Test

For each of the following, mark the letter of the best choice.

**1.** All but two states border each other and make up the main part of the United States. These two states are
- ○ **A** Maine and Vermont.
- ○ **B** Florida and Georgia.
- ○ **C** Texas and New Mexico.
- ○ **D** Alaska and Hawaii.

**2.** Which of the following is the main mountain range located in the East?
- ○ **A** the Rockies
- ○ **B** the Continental Divide
- ○ **C** the Appalachians
- ○ **D** the Cascades

**3.** The Great Lakes are
- ○ **A** the largest freshwater lake system in the world.
- ○ **B** an important waterway for trade with Canada.
- ○ **C** located east of the Mississippi River.
- ○ **D** all of the above.

**4.** Most of the mountains in the Cascades are
- ○ **A** dormant volcanoes.
- ○ **B** under water.
- ○ **C** located in the East.
- ○ **D** more like hills than mountains.

**5.** Climates in the United States
- ○ **A** are mostly hot.
- ○ **B** are mostly cold.
- ○ **C** are similar across the country.
- ○ **D** vary from region to region.

**6.** Oil, natural gas, coal, minerals, and forests are examples of
- ○ **A** pollutants.
- ○ **B** natural resources found in the United States.
- ○ **C** resources the United States lacks.
- ○ **D** natural resources found only in the Northeast.

**7.** Two major seaports in the British colonies were
- ○ **A** New York and Boston.
- ○ **B** Seattle and New York.
- ○ **C** Chicago and Seattle
- ○ **D** Philadelphia and Detroit.

**8.** What type of industry does the Midwest have that is among the most productive of the world?
- ○ **A** coal
- ○ **B** oil
- ○ **C** tourism
- ○ **D** farming

**9.** Pioneers traveled the 2,000 mile Oregon Trail west in search of
- ○ **A** gold, silver, and oil.
- ○ **B** land and oil.
- ○ **C** land and gold.
- ○ **D** oil and religious freedom.

**10.** American movies, television programs, and sports are examples of
- ○ **A** diversity.
- ○ **B** popular culture.
- ○ **C** immigration.
- ○ **D** all of the above.

**11.** The Atlantic Coastal Plain is located in the
- ○ **A** West.
- ○ **B** Midwest.
- ○ **C** Northeast.
- ○ **D** Southeast.

**12.** Which western state is home to more than 10 percent of the U.S. population?
- ○ **A** Texas
- ○ **B** Washington
- ○ **C** California
- ○ **D** Colorado

## Canada



For each of the following, mark the letter of the best choice.

**1.** The main reason for Canada's cold climate is its
- ○ **A** high mountains.
- ○ **B** northern location.
- ○ **C** distance from water.
- ○ **D** lakes and rivers.

**2.** Which part of the country contains most of Canada's mineral deposits?
- ○ **A** the Canadian Shield
- ○ **B** the Grand Banks
- ○ **C** Labrador
- ○ **D** the Pacific Coast

**3.** The building of the transcontinental railroad helped Canada by
- ○ **A** increasing movement to cities.
- ○ **B** linking British Columbia with eastern provinces.
- ○ **C** attracting more immigrants from France.
- ○ **D** giving Native Canadians their own government in Nunavut.

**4.** Which country is Canada's main trading partner?
- ○ **A** China
- ○ **B** Japan
- ○ **C** Great Britain
- ○ **D** the United States

**5.** Which physical feature do the United States and Canada share?
- ○ **A** Niagra Falls
- ○ **B** Rocky Mountains
- ○ **C** St. Lawrence River
- ○ **D** all of the above

**6.** Why do large schools of fish gather at the Grand Banks?
- ○ **A** The cold waters of the Labrador Sea provide an ideal environment.
- ○ **B** The warm waters of the Gulf Stream provide an ideal environment.
- ○ **C** The conditions in the Grand Banks are ideal for growing plankton that fish like to eat.
- ○ **D** Fishers bait the water with food that fish like to eat.

**7.** The area of Canada known for its rainy winters and mild temperatures is

○ **A** British Colombia.
○ **B** the Grand Banks.
○ **C** the Heartland.
○ **D** the Canadian North.

**8.** From which ethnic groups do most Canadians originate?

○ **A** Russian and Chinese
○ **B** British and French
○ **C** Indian and Inuit
○ **D** Italian and German

**9.** Which province was first to have a major Asian population?

○ **A** Alberta
○ **B** British Columbia
○ **C** Manitoba
○ **D** Newfoundland and Labrador

**10.** In what industry do most Canadians work?

○ **A** mining
○ **B** manufacturing
○ **C** trade
○ **D** services

**11.** Which of the following *best* describes the relationship between the United States and Canada?

○ **A** They agree that tariffs are the best way to remain competitive with one another.
○ **B** They share a common history, border, and language.
○ **C** They agreed to ban all beef imports.
○ **D** They are each other's largest trading partner.

**12.** Inuits adapted to the extreme cold of the far north by

○ **A** hunting bison.
○ **B** hunting seals, whales, and walruses.
○ **C** trading with the Vikings.
○ **D** trading with the first Europeans.

Name ______________________ Class ______________ Date ______________

# Mexico

## Chapter 7 Test

For each of the following, mark the letter of the best choice.

**1.** What were Mexico's most valuable resources before oil was discovered?
- ○ **A** coal and natural gas
- ○ **B** scenic landscapes
- ○ **C** rich soils
- ○ **D** minerals

**2.** What is the climate like in Mexico's mountain valleys?
- ○ **A** mild
- ○ **B** hot
- ○ **C** cool
- ○ **D** humid

**3.** What is the vegetation like in northern Mexico?
- ○ **A** desert plants, dry grasslands
- ○ **B** lush, tropical rain forests
- ○ **C** scrub forest
- ○ **D** fertile farmland

**4.** Where did the Olmec live?
- ○ **A** on the Yucatán Peninsula
- ○ **B** on the southern coast of the Gulf of Mexico
- ○ **C** in central Mexico
- ○ **D** on the Isthmus of Tehuantepec

**5.** In what part of Mexico are Indian languages and traditional ways of life most common?
- ○ **A** central
- ○ **B** northern
- ○ **C** southern
- ○ **D** eastern

**6.** Approximately when did the Maya civilization collapse?
- ○ **A** after 1500
- ○ **B** after 900
- ○ **C** before 1500 BC
- ○ **D** after 1300

**7.** To whom did the Spanish monarch grant haciendas?
- ○ **A** favored people of Spanish ancestry
- ○ **B** the Aztecs
- ○ **C** Indian peasants
- ○ **D** mestizos

**8.** A problem in Mexico City that is made worse by the surrounding mountains is

○ **A** corruption.
○ **B** congestion.
○ **C** smog.
○ **D** poverty.

**9.** Which of the following was an achievement of the Maya?

○ **A** They studied the stars and developed a detailed calendar.
○ **B** They built *chinampas.*
○ **C** They conquered neighboring tribes and made them pay taxes.
○ **D** They domesticated an early form of corn.

**10.** Priests taught Indians Spanish and Catholicism at church outposts known as

○ **A** haciendas.
○ **B** chinampas.
○ **C** maquiladoras.
○ **D** missions.

**11.** Most Mexicans speak Spanish because

○ **A** of Spanish influence in colonial times.
○ **B** they used to live in Spain.
○ **C** the Indian languages and Spanish are very similar.
○ **D** about 90 percent of Mexican citizens were born in Spain.

**12.** Mexico's coastal resorts and Aztec and Maya monuments attract many

○ **A** immigrants.
○ **B** tourists.
○ **C** farmers.
○ **D** miners.

## Central America and the Caribbean Chapter 8 Test

For each of the following, mark the letter of the best choice.

**1.** Which island group lies in the Atlantic Ocean southeast of Florida?
- ○ **A** Lesser Antilles
- ○ **B** Greater Antilles
- ○ **C** Bahamas
- ○ **D** Hispaniola

**2.** Economic development of the region is limited because it
- ○ **A** lies in an earthquake zone.
- ○ **B** is threatened by hurricanes.
- ○ **C** depends on sugarcane.
- ○ **D** has few energy resources.

**3.** Cloud forests are found
- ○ **A** in highland regions.
- ○ **B** on coastal plains.
- ○ **C** in savannas.
- ○ **D** along rivers.

**4.** An export of rain forests in the region is
- ○ **A** coffee.
- ○ **B** cotton.
- ○ **C** timber.
- ○ **D** sugarcane.

**5.** The United Fruit Company controlled most of the banana production in Central America in
- ○ **A** the early to mid-1900s.
- ○ **B** the 1800s.
- ○ **C** early colonial times.
- ○ **D** the mid- to late 1900s.

**6.** What is the official language in most Central American countries?
- ○ **A** English
- ○ **B** Spanish
- ○ **C** French
- ○ **D** Creole

**7.** Some traditional foods in Central America are the same as in
- ○ **A** Mexico.
- ○ **B** Spain.
- ○ **C** England.
- ○ **D** France.

**8.** Where in Central America do people of African descent most often live?
- ○ **A** in mountain valleys.
- ○ **B** in the highlands.
- ○ **C** in the rain forests.
- ○ **D** along the Caribbean coast.

**9.** Two Central American countries with little land for agriculture are
- ○ **A** Guatemala and Costa Rica.
- ○ **B** Belize and Honduras.
- ○ **C** El Salvador and Panama.
- ○ **D** Nicaragua and Guatemala.

**Study the map below and answer the question that follows.**

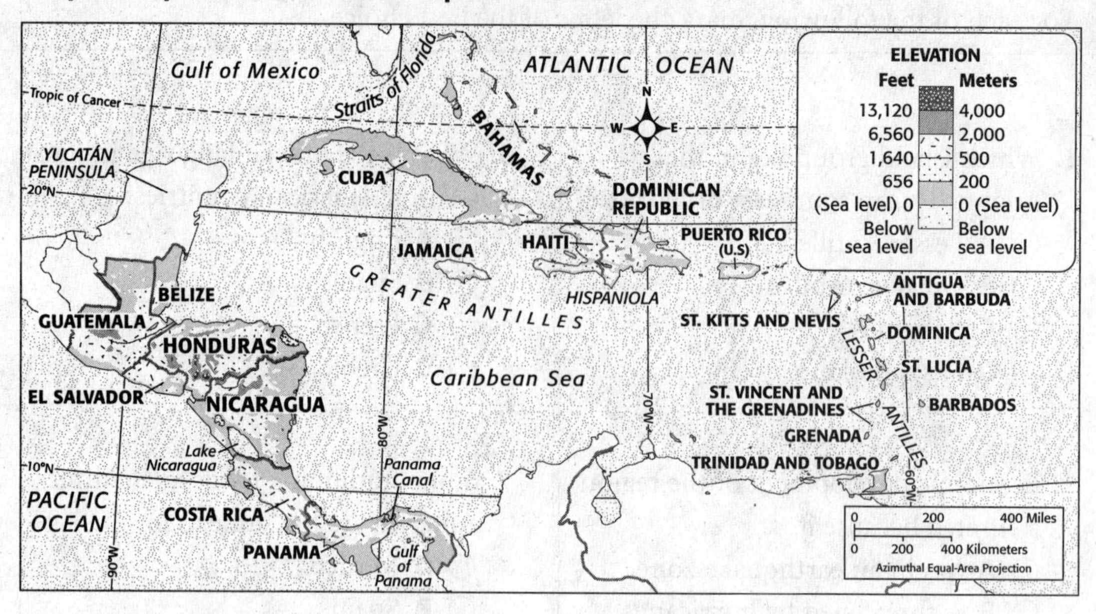


**10.** The island group closest to South America is the

- ○ **A** Greater Antilles.
- ○ **B** Dominican Republic.
- ○ **C** Lesser Antilles.
- ○ **D** Bahamas.

**11.** Most people in El Salvador live in poverty because

- ○ **A** the country has little fertile soil.
- ○ **B** a few rich people own most of the land.
- ○ **C** a civil war is being fought there.
- ○ **D** the tourism industry provides most jobs.

**12.** Which Central American country has a history of peace and stability?

- ○ **A** Honduras
- ○ **B** Guatemala
- ○ **C** Costa Rica
- ○ **D** Nicaragua

Name ______________________ Class ____________ Date ____________



For each of the following, mark the letter of the best choice.

**1.** What country was the home of the Chibcha before the Spanish arrived?
- ○ **A** Venezuela
- ○ **B** Suriname
- ○ **C** Guyana
- ○ **D** Colombia

**2.** The countries in Caribbean South America that export oil are
- ○ **A** Venezuela and Guyana.
- ○ **B** Colombia and Suriname.
- ○ **C** Venezuela and Colombia.
- ○ **D** Colombia and Guyana.

**3.** In what year did Venezuela officially become independent of Spain?
- ○ **A** 1789
- ○ **B** 1811
- ○ **C** 1830
- ○ **D** 1960

**4.** In Venezuela, the joropo is a
- ○ **A** traditional Indian ring toss game.
- ○ **B** lively couples' dance.
- ○ **C** type of rodeo.
- ○ **D** form of political protest.

**5.** The legendary land, rich in gold, inspired by the Chibcha culture is called
- ○ **A** Cartagena.
- ○ **B** New Granada.
- ○ **C** El Dorado.
- ○ **D** Paramaribo.

**6.** Which European country colonized Suriname?
- ○ **A** the Netherlands
- ○ **B** Great Britain
- ○ **C** France
- ○ **D** Spain

**7.** Which crop does the government of Colombia want to export less of?
- ○ **A** flowers
- ○ **B** sugarcane
- ○ **C** coca
- ○ **D** coffee

**8.** In what region of Caribbean South America do people of African descent mostly live?

○ **A** tropical rain forest
○ **B** high mountains
○ **C** coastal areas
○ **D** grassland plains

**9.** Which country is a member of the Organization of Petroleum Exporting Countries (OPEC)?

○ **A** Venezuela
○ **B** Colombia
○ **C** French Guiana
○ **D** Suriname

**10.** Which country does *not* include part of the Guiana Highlands?

○ **A** Suriname
○ **B** Colombia
○ **C** Venezuela
○ **D** Guyana

**11.** Angel Falls in Venezuela is the world's

○ **A** oldest waterfall.
○ **B** widest waterfall.
○ **C** highest waterfall.
○ **D** most visited waterfall.

**12.** Which of the following was the Chibcha culture *not* known for?

○ **A** fine gold objects
○ **B** growing coffee
○ **C** weaving
○ **D** pottery making

Name ______________________ Class ______________ Date ______________

## Atlantic South America

# Chapter 10 Test

For each of the following, mark the letter of the best choice.

**1.** The southernmost country in Atlantic South America is

- ○ **A** Brazil.
- ○ **B** Argentina.
- ○ **C** Uruguay.
- ○ **D** Paraguay.

**2.** The world's largest river system is the

- ○ **A** Río de la Plata.
- ○ **B** Amazon.
- ○ **C** Uruguay.
- ○ **D** Paraná.

**3.** The wide, grassy plains in central Argentina are called

- ○ **A** the Pampas.
- ○ **B** the Andes.
- ○ **C** plateaus.
- ○ **D** rain forests.

**4.** The largest country in South America is

- ○ **A** Argentina.
- ○ **B** Uruguay.
- ○ **C** Paraguay.
- ○ **D** Brazil.

**5.** The first people who lived in Brazil were

- ○ **A** Spanish settlers.
- ○ **B** Portuguese settlers.
- ○ **C** French explorers.
- ○ **D** American Indians.

**6.** A megacity is

- ○ **A** a large city.
- ○ **B** a giant urban area that includes surrounding cities and suburbs.
- ○ **C** two cities in one.
- ○ **D** a city with more than one million people in it.

**7.** Argentina's informal economy is based on

- ○ **A** odd jobs.
- ○ **B** ranching.
- ○ **C** farming.
- ○ **D** education.

**8.** The only landlocked country in Atlantic South America is

○ **A** Brazil.
○ **B** Paraguay.
○ **C** Uruguay.
○ **D** Argentina.

**9.** Brazil's two largest cities are

○ **A** Asunción and Montevideo.
○ **B** Brasília and Mercosur.
○ **C** São Paulo and Rio de Janeiro.
○ **D** Manaus and Buenos Aires.

**10.** Which country has the largest population of Roman Catholics in the world?

○ **A** Brazil
○ **B** Argentina
○ **C** Uruguay
○ **D** Paraguay

**11.** The Amazon Basin is a

○ **A** rugged plateau.
○ **B** giant floodplain.
○ **C** huge estuary.
○ **D** dry desert.

**12.** One aspect of Brazilian life that reflects its mix of cultures is

○ **A** that more than 90 percent of Brazilians live in urban areas.
○ **B** that almost all Brazilians speak both Spanish and Guarani.
○ **C** the celebration of Carnival.
○ **D** its large economy.

Name ______________________ Class ______________ Date ______________



For each of the following, mark the letter of the best choice.

**1.** Which of the following is a characteristic of Chile's economy?

○ **A** Poverty rates are increasing.
○ **B** The climate does not allow farmers to grow crops.
○ **C** International trade is key.
○ **D** Tourism forms the basis of Chile's economy.

**2.** What is the term for the broad, high plateau that lies between the ridges of the Andes?

○ **A** strait
○ **B** altiplano
○ **C** El Niño
○ **D** Atacama

**3.** Which of the following was a terrorist group active in Peru in the 1980s and 1990s?

○ **A** the Young Towns
○ **B** the coup
○ **C** the Shining Path
○ **D** the Creoles

**4.** In colonial times the Spanish viceroy

○ **A** ruled jointly with the Inca king.
○ **B** shared the wealth from the gold mines with the native Indians.
○ **C** was elected by Spaniards in Peru.
○ **D** made sure the Indians followed Spanish laws and customs.

**5.** Right after the coup in the 1970s, Chile

○ **A** elected a president who had some ideas influenced by communism.
○ **B** created a new, democratic government.
○ **C** was ruled by a harsh military government.
○ **D** focused on improving human rights.

**6.** The Galápagos Islands

- ○ **A** have wildlife found nowhere else in the world.
- ○ **B** receive rain only about five times per century.
- ○ **C** are separated from South America by the Strait of Magellan.
- ○ **D** are cold year round, despite being close to the equator.

**7.** Which country has two capital cities?

- ○ **A** Ecuador
- ○ **B** Bolivia
- ○ **C** Peru
- ○ **D** Chile

**8.** The large island that is south of the Strait of Magellan is called

- ○ **A** Paraguay.
- ○ **B** the alitplano.
- ○ **C** La Paz.
- ○ **D** Tierra del Fuego.

**9.** Which of the following is an effect of El Niño?

- ○ **A** Fish swarm to affected areas.
- ○ **B** The cool water of the Pacific Ocean warms up.
- ○ **C** Rivers cut through dry coastal regions.
- ○ **D** Areas along the coast experience drought.

**10.** The Atacama Desert is known for being very dry and

- ○ **A** sunny.
- ○ **B** hot.
- ○ **C** cloudy.
- ○ **D** windy.

**11.** The Inca Empire was once home to as many as

- ○ **A** 15,000 people.
- ○ **B** 300,000 people.
- ○ **C** 4 million people.
- ○ **D** 12 million people.

**12.** As advanced as the Inca civilization was, it did not have

- ○ **A** roads.
- ○ **B** a written language.
- ○ **C** bridges.
- ○ **D** irrigation systems.

Name ________________ Class ________ Date ________

# Southern Europe

## Chapter 12 Test

For each of the following, mark the letter of the best choice.

**1.** What are two central elements of Greek culture?
- ○ **A** church and family
- ○ **B** family and politics
- ○ **C** politics and church
- ○ **D** history and politics

**2.** In the 1800s, Greece became a monarchy after
- ○ **A** holding free elections.
- ○ **B** losing their colonies.
- ○ **C** overthrowing the Turks.
- ○ **D** conquering Spain.

**3.** The head of the Roman Catholic Church lives in
- ○ **A** Venice.
- ○ **B** Rome.
- ○ **C** Vatican City.
- ○ **D** Sicily.

**4.** Why is southern Italy poorer than northern Italy?
- ○ **A** It has less industry.
- ○ **B** It has no tourism.
- ○ **C** It has no coastal ports.
- ○ **D** It has rich, fertile farmland.

**5.** What is the westernmost peninsula in Europe?
- ○ **A** Iberia
- ○ **B** Balkan
- ○ **C** Italy
- ○ **D** Pindus

**6.** Architecture in Spain is influenced by
- ○ **A** traditional fados.
- ○ **B** Spain's Muslim past.
- ○ **C** the Italians.
- ○ **D** Spain's colonies.

**7.** Two important crops in Portugal are
- ○ **A** grapes and cork.
- ○ **B** pineapples and tomatoes.
- ○ **C** tomatoes and cork.
- ○ **D** grapes and pineapples.

**8.** What are the highest mountains in Europe?

- ○ **A** the Apennines
- ○ **B** the Pindus
- ○ **C** the Alps
- ○ **D** the Pyrenees

**9.** What form of government returned to Greece in 1974?

- ○ **A** monarchy
- ○ **B** democracy
- ○ **C** dictatorship
- ○ **D** republic

**10.** What influenced the beginning of the Renaissance?

- ○ **A** Rich merchants supported artists.
- ○ **B** Nationalism swept through Italy.
- ○ **C** Rome ruled a vast empire.
- ○ **D** Italians overthrew Muslim rulers.

**11.** The people of Spain and Portugal are almost entirely

- ○ **A** Eastern Orthodox.
- ○ **B** Muslim.
- ○ **C** Protestant.
- ○ **D** Roman Catholic.

**12.** Which river runs through one of the most fertile and densely populated areas in Italy?

- ○ **A** the Ebro
- ○ **B** the Po
- ○ **C** the Douro
- ○ **D** the Tagus

Name ______________________ Class ______________ Date ____________



For each of the following, mark the letter of the best choice.

**1.** What landform stretches from the Atlantic coast into Eastern Europe?
- ○ **A** Alps
- ○ **B** Jura Mountains
- ○ **C** Northern European Plain
- ○ **D** Balkan Peninsula

**2.** The leading agricultural producer in the European Union is
- ○ **A** France.
- ○ **B** Germany.
- ○ **C** Austria.
- ○ **D** the Netherlands.

**3.** Germany's economy is based on
- ○ **A** banking.
- ○ **B** industry.
- ○ **C** tourism.
- ○ **D** farming.

**4.** What landform covers much of Switzerland and Austria?
- ○ **A** plains
- ○ **B** uplands
- ○ **C** plateau
- ○ **D** mountains

**5.** West-Central Europe's marine west coast climate is a key natural resource for
- ○ **A** raising livestock.
- ○ **B** farming.
- ○ **C** fishing.
- ○ **D** industry.

**6.** During World War II, France and the Benelux Countries were invaded by
- ○ **A** Austria.
- ○ **B** Germany.
- ○ **C** Poland.
- ○ **D** Russia.

**7.** The Benelux Countries are all
- ○ **A** landlocked.
- ○ **B** centers for farming.
- ○ **C** densely populated.
- ○ **D** Communist.

**8.** The Dutch have used dikes and other technology to
- ○ **A** raise the sea level.
- ○ **B** create hydroelectric power.
- ○ **C** create polders.
- ○ **D** improve communication.

**Study the map below and answer the question that follows.**

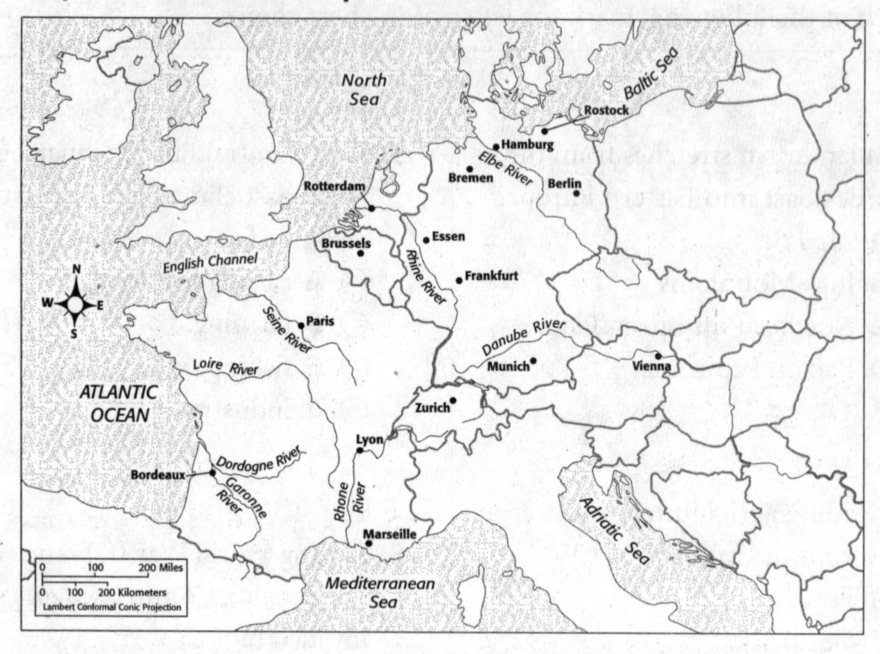


**9.** Which river runs through Vienna?

- ○ **A** Rhine River
- ○ **B** Danube River
- ○ **C** Elbe River
- ○ **D** Oder River

**10.** Which of the following is a highly urbanized and industrialized area in the Netherlands?

- ○ **A** Brussels
- ○ **B** Hamburg
- ○ **C** Rotterdam
- ○ **D** Bern

**11.** What industry employs more than half of Austria's workforce?

- ○ **A** trade
- ○ **B** mining
- ○ **C** service
- ○ **D** manufacturing

**12.** Belgium has been troubled by regional conflicts over

- ○ **A** trade and economic issues.
- ○ **B** democratic government.
- ○ **C** immigration.
- ○ **D** language and other cultural issues.

Name ______________________ Class ______________ Date ______________

## Northen Europe



For each of the following, mark the letter of the best choice.

**1.** The physical geography of Northern Europe
- ○ **A** changes dramatically in different locations.
- ○ **B** is similar throughout the region.
- ○ **C** is similar to Southern Europe.
- ○ **D** is similar to West-Central Europe.

**2.** Which country uses about 50 percent of its land for farming?
- ○ **A** Finland
- ○ **B** Norway
- ○ **C** Denmark
- ○ **D** Sweden

**3.** The seas and oceans of Northern Europe
- ○ **A** do not contain natural resources.
- ○ **B** keep the climates cold.
- ○ **C** have provided fish to the people for centuries.
- ○ **D** provide geothermal energy to Iceland.

**4.** In spite of the region's latitude, it's climates are
- ○ **A** surprisingly cold.
- ○ **B** surprisingly mild.
- ○ **C** extremely dry.
- ○ **D** uninhabitable.

**5.** Which of the following make up the United Kingdom?
- ○ **A** England, Wales, Iceland, Northern Ireland
- ○ **B** England, Scotland, Wales, Northern Ireland
- ○ **C** Iceland, England, Greenland, Northern Ireland
- ○ **D** England, Greenland, Wales, Northern Ireland

**6.** The Industrial Revolution helped
- ○ **A** the Vikings gain colonies.
- ○ **B** Ireland win its independence.
- ○ **C** the United Kingdom build a large empire.
- ○ **D** the United Kingdom limit worldwide trade.

**7.** The governments of Ireland and the United Kingdom are
- ○ **A** democracies.
- ○ **B** monarchies.
- ○ **C** dictatorships.
- ○ **D** Communist.

**8.** The conflict in Northern Ireland
- ○ **A** was finally resolved in the 1990s.
- ○ **B** never affected the United Kingdom.
- ○ **C** is between Catholics and Protestants.
- ○ **D** is between judges and elected officials.

**9.** Viking warriors
- ○ **A** formed an excellent infantry and attacked over land.
- ○ **B** fought many battles but did not win many.
- ○ **C** were not powerful until the 1500s.
- ○ **D** were excellent sailors and attacked by sea.

**10.** All of the Scandinavian countries
- ○ **A** have strong economies and high standards of living.
- ○ **B** have refused to join the European Union.
- ○ **C** use geothermal energy for heating homes.
- ○ **D** lag far behind in the citizens' education.

**11.** Greenland is
- ○ **A** an independent country.
- ○ **B** a territory of Denmark.
- ○ **C** an important farming area.
- ○ **D** an enemy of the United Kingdom.

**12.** Tourists often come to Iceland to see
- ○ **A** fjords and emerald green hills.
- ○ **B** historical monuments.
- ○ **C** pine forests and lumber mills.
- ○ **D** volcanoes, glaciers, and geysers.

Name ______________________ Class ______________ Date ____________

# Eastern Europe

# Chapter 15 Test

For each of the following, mark the letter of the best choice.

**1.** Most of the Balkan Peninsula is covered by
- ○ **A** lakes.
- ○ **B** grasslands.
- ○ **C** marshes.
- ○ **D** mountains.

**2.** What body of water is most important to trade and transportation throughout Eastern Europe?
- ○ **A** Rhine
- ○ **B** Danube
- ○ **C** Vistula
- ○ **D** Aegean Sea

**3.** In 1986 a disaster at Chernobyl that ruined forests and soil was caused by
- ○ **A** terrorists.
- ○ **B** massive flooding.
- ○ **C** a nuclear power plant explosion.
- ○ **D** polluted rivers.

**4.** After World War II much of Eastern Europe was forced to accept governments run or influenced by
- ○ **A** the Baltic Republics.
- ○ **B** Germany.
- ○ **C** Serbia.
- ○ **D** the Soviet Union.

**5.** Which of the following is a Baltic republic?
- ○ **A** Belarus
- ○ **B** Estonia
- ○ **C** Budapest
- ○ **D** Albania

**6.** After the collapse of the Soviet Union in 1989, the governments of many Eastern European countries became
- ○ **A** EU members.
- ○ **B** democratic.
- ○ **C** Communist.
- ○ **D** CIS members.

**7.** Some of the biggest differences among the cultures of Poland and the Baltic Republics are in their
- ○ **A** embroidery styles.
- ○ **B** crafts.
- ○ **C** languages and religions.
- ○ **D** food.

**8.** After the fall of the Soviet Union, the economies of many Eastern European countries were
- ○ **A** much weaker than in Western European countries.
- ○ **B** much stronger than in Western European countries.
- ○ **C** about the same as in Western European countries.
- ○ **D** ahead of Western Europe because of tourism.

**9.** Which of these Eastern European countries split into two countries in 1993?
- ○ **A** Yugoslavia
- ○ **B** Moldavia
- ○ **C** Czechoslovakia
- ○ **D** Bosnia and Herzegovina

**10.** One of the most prosperous cities of inland Eastern Europe is
- ○ **A** Warsaw.
- ○ **B** Tallinn.
- ○ **C** Budapest.
- ○ **D** Mostar.

**11.** Much of the violence that occurred in the Balkans in the 1990s resulted from
- ○ **A** Soviet domination.
- ○ **B** poor economic planning.
- ○ **C** ethnic conflict.
- ○ **D** mistreatment of Magyars.

**12.** The countries of Serbia and Montenegro and Croatia were formed
- ○ **A** after the murder by a Serbian of the heir to the Austro-Hungarian throne.
- ○ **B** after World War I ended.
- ○ **C** after the breakup of Czechoslovakia.
- ○ **D** after the breakup of Yugoslavia.

Name ______________________ Class ____________ Date ____________



For each of the following, mark the letter of the best choice.

**1.** What two continents meet in Russia's Ural Mountains?
- ○ **A** Europe and Antarctica
- ○ **B** Asia and Africa
- ○ **C** Europe and Asia
- ○ **D** Europe and Africa

**2.** The Caucasus Mountains form a border between Russia and
- ○ **A** Georgia and Azerbaijan.
- ○ **B** Armenia and Ukraine.
- ○ **C** Georgia and Armenia.
- ○ **D** Armenia and Azerbaijan.

**3.** Russia's capital and largest city is
- ○ **A** Tbilisi.
- ○ **B** Baku.
- ○ **C** Kiev.
- ○ **D** Moscow.

**4.** Siberia is best described as a
- ○ **A** small region with a tropical climate.
- ○ **B** vast area with a harsh climate.
- ○ **C** flat, marshy land with a mild climate.
- ○ **D** barren plain with a dry, desert climate.

**5.** The Caucasus region has lowlands along the shores of the
- ○ **A** Arctic Ocean and Pacific Ocean.
- ○ **B** Sea of Okhotsk and Caspian Sea.
- ○ **C** Black Sea and Caspian Sea.
- ○ **D** Black Sea and Baltic Sea.

**6.** As czar, Ivan IV was such a cruel and savage ruler he became known as
- ○ **A** Ivan the Paranoid.
- ○ **B** Ivan the Tyrant.
- ○ **C** Ivan the Terrible.
- ○ **D** Prince of Muscovy.

**7.** The radical Communist group that took over Russia in the Russian Revolution was the
- ○ **A** Vikings.
- ○ **B** Slavs.
- ○ **C** Mongols.
- ○ **D** Bolsheviks.

**8.** During the fall of the Soviet Union, its leader was
- ○ **A** Vladimir Lenin.
- ○ **B** Joseph Stalin.
- ○ **C** Peter the Great.
- ○ **D** Mikhail Gorbachev.

**9.** The Russian Federation is governed by an elected president and a legislature called the
- ○ **A** Federal Assembly.
- ○ **B** Soviets.
- ○ **C** Senate.
- ○ **D** Bolsheviks.

**10.** Of the countries in the Caucasus Region, the only one that is landlocked is
- ○ **A** Russia.
- ○ **B** Georgia.
- ○ **C** Armenia.
- ○ **D** Azerbaijan.

**11.** What government body is appointed to run each government of the Caucasus countries today?
- ○ **A** parliament
- ○ **B** president
- ○ **C** prime minister
- ○ **D** czar

**12.** The vast forest that covers much of Russia is called the
- ○ **A** taiga.
- ○ **B** steppe.
- ○ **C** tundra.
- ○ **D** uplands.

Name ______________________ Class ______________ Date ____________

# The Eastern Mediterranean

## Chapter 17 Test

For each of the following, mark the letter of the best choice.

**1.** The Eastern Mediterranean lies between the continents of
- ○ **A** South America and North America.
- ○ **B** Asia and Australia.
- ○ **C** Europe and Asia.
- ○ **D** Africa and Australia.

**2.** The Jordan River empties into the
- ○ **A** Black Sea.
- ○ **B** Mediterranean Sea.
- ○ **C** Red Sea.
- ○ **D** Dead Sea.

**3.** Much of Syria and Jordan are covered by the
- ○ **A** Syrian Desert.
- ○ **B** Negev.
- ○ **C** Pontic Mountains.
- ○ **D** Taurus Mountains.

**4.** Success of commercial farming in the region relies heavily on
- ○ **A** irrigation.
- ○ **B** foreign aid.
- ○ **C** desert conditions.
- ○ **D** nomadic herding.

**5.** When the Romans invaded the area that is now called Turkey, they captured the strategic city of Byzantium and renamed it
- ○ **A** Damascus.
- ○ **B** Rome.
- ○ **C** Constantinople.
- ○ **D** Tripoli.

**6.** What was the name of the series of invasions of Palestine launched by Christians from Europe?
- ○ **A** the Diaspora
- ○ **B** the Crusades
- ○ **C** Zionism
- ○ **D** Knesset

**7.** Which country shares a border with both the Mediterranean and the Black seas?
- ○ **A** Turkey
- ○ **B** Israel
- ○ **C** Lebanon
- ○ **D** Syria

**8.** Which country has no coastline on the Mediterranean Sea?
- ○ **A** Turkey
- ○ **B** Syria
- ○ **C** Israel
- ○ **D** Jordan

**9.** Where does the Jordan River begin?
- ○ **A** Turkey
- ○ **B** Syria
- ○ **C** Israel
- ○ **D** Jordan

**10.** What are phosphates used to make?
- ○ **A** mercury
- ○ **B** concrete
- ○ **C** fertilizers
- ○ **D** tar

**11.** When was Palestine declared to be the nation of Israel?
- ○ **A** the late 1800s
- ○ **B** 1916
- ○ **C** 1948
- ○ **D** 1980

**12.** Who controlled Palestine after World War I?
- ○ **A** Arabs
- ○ **B** Ottoman Turks
- ○ **C** Bedouins
- ○ **D** the British

Name ______________________ Class ______________ Date ____________

# The Arabian Peninsula, Iraq, and Iran — Chapter 18 Test

For each of the following, mark the letter of the best choice.

**1.** Which of the following is found on the Arabian Peninsula?
- ○ **A** the largest sand desert in the world
- ○ **B** the largest mountain range in the world
- ○ **C** permanent rivers
- ○ **D** tropical forests

**2.** Which country in the region is rich in mineral deposits?
- ○ **A** Iraq
- ○ **B** Iran
- ○ **C** Bahrain
- ○ **D** Qatar

**3.** The poorest country in the region is
- ○ **A** Yemen.
- ○ **B** Oman.
- ○ **C** Iraq.
- ○ **D** Iran.

**4.** The Persian Gulf War began after
- ○ **A** Saddam Hussein invaded Iran.
- ○ **B** Saddam Hussein invaded Kuwait.
- ○ **C** weapons of mass destruction were found in Iraq.
- ○ **D** the terrorist attacks of September 11, 2001.

**5.** Which country in the region is mostly covered with plateaus and mountains?
- ○ **A** Iraq
- ○ **B** Iran
- ○ **C** Yemen
- ○ **D** Oman

**6.** The name of the *Rub' al-Khali* desert in Saudi Arabia means
- ○ **A** land of riches.
- ○ **B** land between the rivers.
- ○ **C** Fertile Crescent.
- ○ **D** Empty Quarter.

**7.** The desert plains in the northern part of the Arabian Peninsula are covered with
- ○ **A** sand.
- ○ **B** wadis.
- ○ **C** volcanic rock.
- ○ **D** coral reefs.

**8.** The world's first civilization was located in present-day
- ○ **A** Iran.
- ○ **B** Iraq.
- ○ **C** Saudi Arabia.
- ○ **D** Oman.

**9.** Which is true of the region's climate?
- ○ **A** It never snows.
- ○ **B** Temperatures are always hot.
- ○ **C** Days are hot and nights are cold.
- ○ **D** Some mountain peaks receive only 4 inches of rain per year.

**10.** Which country is made of seven tiny kingdoms?
- ○ **A** Oman
- ○ **B** Yemen
- ○ **C** Bahrain
- ○ **D** United Arab Emirates

**11.** One influence of Islam on the region is that
- ○ **A** boys and girls go to school together.
- ○ **B** many women own and run businesses.
- ○ **C** men and women wear clothes that cover their arms and legs.
- ○ **D** women often appear in public alone.

**12.** Who has ruled Saudi Arabia since 1932?
- ○ **A** ayatollahs
- ○ **B** Persians
- ○ **C** an elected legislature
- ○ **D** members of the Saud family

Name ______________________ Class ______________ Date ____________

# Central Asia

# Chapter 19 Test

For each of the following, mark the letter of the best choice.

**1.** Which word *best* describes Central Asia's physical geography?
- ○ **A** coastal
- ○ **B** fertile
- ○ **C** landlocked
- ○ **D** arable

**2.** What physical features have contributed to the isolation of the Central Asia region?
- ○ **A** fertile plains
- ○ **B** rugged mountains
- ○ **C** nomadic lifestyle
- ○ **D** the Silk Road

**3.** What is the relationship between climate and vegetation in Central Asia?
- ○ **A** A humid, rainy climate produces many fertile valleys and varied plants.
- ○ **B** The region's vegetation is not affected by climate.
- ○ **C** A harsh, dry climate makes it hard for plants to grow.
- ○ **D** The mild, desert climate produces most of the region's crops.

**4.** What best explains why vegetation does not grow in the region's mountain peaks?
- ○ **A** Peaks are too high to get to for planting crops.
- ○ **B** Animals that graze there have eaten all the plants.
- ○ **C** There is too much rain at high elevations.
- ○ **D** It is too cold, dry, and windy.

**5.** What is the reason the Aral Sea is shrinking?
- ○ **A** canal building
- ○ **B** irrigation
- ○ **C** overfishing
- ○ **D** tourism

**6.** What is preventing Central Asia from exporting oil?
- ○ **A** It needs to build pipelines through rugged mountains.
- ○ **B** It has too little oil.
- ○ **C** Its ocean ports are polluted.
- ○ **D** The people who live there use all the oil.

**7.** Which country shares a large border with Russia?
- ○ **A** Afghanistan
- ○ **B** Turkmenistan
- ○ **C** Uzbekistan
- ○ **D** Kazakhstan

**8.** What are some key natural resources in Central Asia?
- ○ **A** water, oil and gas, other minerals
- ○ **B** water, oil and gas, cotton
- ○ **C** silver, coal, rain forest
- ○ **D** mountains, deserts, oceans

**9.** Who was the last major group to rule Central Asia?
- ○ **A** Turkic nomads
- ○ **B** Arabs
- ○ **C** Mongols
- ○ **D** Soviets

**10.** Which situation in Central Asia shows Arab influence?
- ○ **A** Yurts are an important symbol of the region's heritage.
- ○ **B** Islam is the main religion.
- ○ **C** The region is now a group of independent republics.
- ○ **D** Many people are Russian Orthodox.

**11.** What is a yurt?
- ○ **A** large apartment building in a city
- ○ **B** moveable home used by nomads
- ○ **C** traditional religious building
- ○ **D** mountain village

**12.** Why do people speak different languages in Central Asia?
- ○ **A** Each government established an official language.
- ○ **B** Each religion requires its people to speak a certain language.
- ○ **C** Each ethnic group speaks its own language.
- ○ **D** Russian rulers encouraged people to continue using their traditional languages.

Name ______________________ Class ______________ Date ______________

## North Africa

## Chapter 20 Test

For each of the following, mark the letter of the best choice.

**1.** What covers most of North Africa?
- ○ **A** fertile wetlands
- ○ **B** the Sahara
- ○ **C** the Atlas Mountains
- ○ **D** spring-fed oases

**2.** An important link between the Mediterranean and Red seas is the
- ○ **A** Sinai Peninsula.
- ○ **B** Strait of Gibraltar.
- ○ **C** Suez Canal.
- ○ **D** Nile River.

**3.** What type of climate does most of North Africa have?
- ○ **A** desert
- ○ **B** Mediterranean
- ○ **C** steppe
- ○ **D** tropical

**4.** Ancient Egyptians buried their kings in
- ○ **A** unmarked graves.
- ○ **B** marble temples.
- ○ **C** small villages.
- ○ **D** stone pyramids.

**5.** The ancient Egyptians developed a sophisticated writing system called
- ○ **A** hieroglyphics.
- ○ **B** emblems.
- ○ **C** Egyptology.
- ○ **D** rendering.

**6.** When did Arab powers from Southwest Asia rule North Africa?
- ○ **A** during the 2500s BC
- ○ **B** from the AD 600s to the 1800s
- ○ **C** during the 1900s
- ○ **D** from the 1800s to today

**7.** What religion do most North Africans practice?
- ○ **A** Christianity
- ○ **B** Islam
- ○ **C** Judaism
- ○ **D** Buddhism

**8.** What is Ramadan?
- ○ **A** a holy month during which Muslims do not eat or drink during the day
- ○ **B** Muhammad's birthday which is marked with lights and parades
- ○ **C** a city on Morocco's Mediterranean coast
- ○ **D** an ancient Egyptian king who brought Islam to the Maghreb

**Study the map below and answer the question that follows**

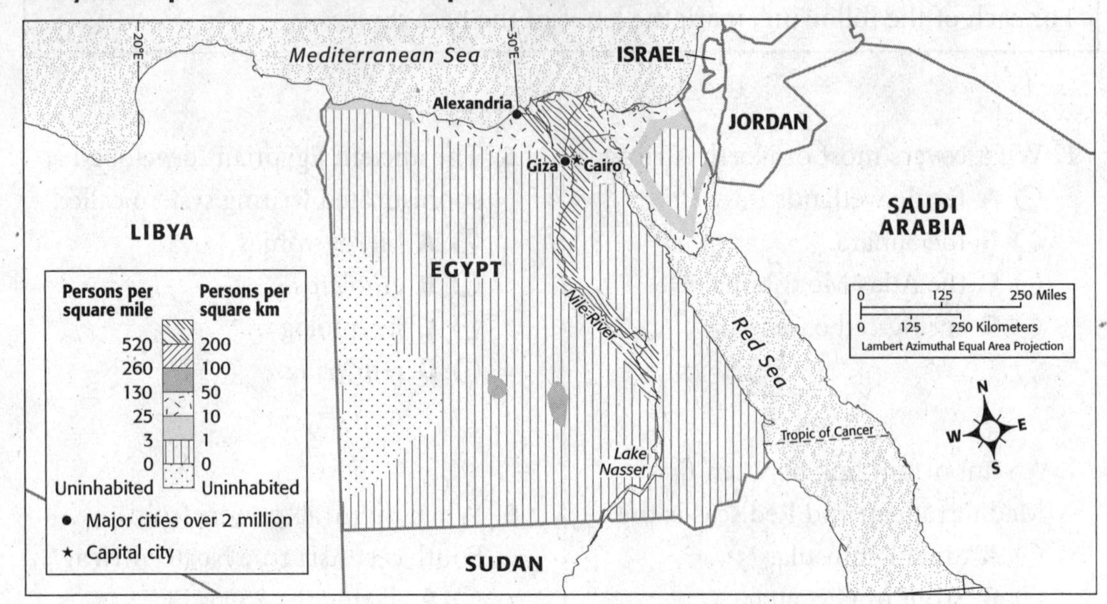


**9.** Which region of Egypt appears to have no population? Why?

- ○ **A** northeastern Egypt, because it is usually flooded
- ○ **B** northern Egypt, because it contains rocky mountains
- ○ **C** southeastern Egypt, because Lake Nassar occupies most of it
- ○ **D** western Egypt, because barren desert covers it

**10.** People in North Africa often greet each other by

- ○ **A** hugging and linking arms.
- ○ **B** bowing from the waist.
- ○ **C** shaking hands and touching their hands to their hearts.
- ○ **D** waving their arms over their heads.

**11.** Most rural Egyptians are farmers called

- ○ **A** Maghreb.
- ○ **B** fellahin.
- ○ **C** Berbers.
- ○ **D** Saharans.

**12.** What is Egypt's second-largest city?

- ○ **A** Alexandria
- ○ **B** Tunis
- ○ **C** Tripoli
- ○ **D** Cairo

# West Africa

## Chapter 21 Test

For each of the following, mark the letter of the best choice.

**1.** The climate of West Africa
- ○ **A** changes significantly from west to east.
- ○ **B** changes significantly from north to south.
- ○ **C** is consistent throughout the region.
- ○ **D** changes most according to elevation.

**2.** The inland delta of the Niger River
- ○ **A** extends along the coast of Liberia.
- ○ **B** prevents the river from emptying into the ocean.
- ○ **C** is hundreds of miles from the coast.
- ○ **D** prevents people from fishing in the river.

**3.** The desert climate extends across the region's
- ○ **A** northern zone.
- ○ **B** eastern zone.
- ○ **C** western zone.
- ○ **D** southern zone.

**4.** The steppe climate of the Sahel supports
- ○ **A** logging in the forests.
- ○ **B** farming of staple crops.
- ○ **C** fruit orchards.
- ○ **D** grazing animals.

**5.** Which of the following is an important crop in West Africa?
- ○ **A** rice
- ○ **B** cacao
- ○ **C** corn
- ○ **D** sugarcane

**6.** The great kingdoms of West Africa became powerful by
- ○ **A** conquering Mediterranean countries.
- ○ **B** trading with groups in East Africa.
- ○ **C** using ships to send goods around the world.
- ○ **D** controlling trade routes across the Sahara.

**7.** The Atlantic slave trade

- ○ **A** supplied labor for European colonies.
- ○ **B** supplied slaves to North Africa.
- ○ **C** was opposed by most European countries.
- ○ **D** resulted in many Africans being shipped to Europe.

**8.** Political boundaries drawn by European colonizers

- ○ **A** often separated members of the same ethnic group.
- ○ **B** kept people who spoke the same language together.
- ○ **C** kept people of the same religion together.
- ○ **D** generally followed climate zones.

**9.** Clothing in West Africa is loose and flowing because of

- ○ **A** the shortage of cloth.
- ○ **B** traditional religious beliefs.
- ○ **C** the influence of Western styles.
- ○ **D** the warm climate.

**10.** Which of the following statements best describes Nigeria?

- ○ **A** Nigeria is rich in resources, but many Nigerians are poor.
- ○ **B** Nigeria has few resources, and its people are poor.
- ○ **C** Nigeria is rich in resources, and most Nigerians have a high standard of living.
- ○ **D** Nigeria has few resources but is able to provide food for its people.

**11.** Many coastal countries in West Africa

- ○ **A** have large oil reserves.
- ○ **B** have had stable, democratic governments.
- ○ **C** depend on farming for their economies.
- ○ **D** have joined together in a political union.

**12.** Drought has been a significant problem for

- ○ **A** Nigeria.
- ○ **B** the coastal countries.
- ○ **C** the Sahel countries.
- ○ **D** all the countries in West Africa.

Name ____________________ Class ____________ Date ____________

# East Africa

# Chapter 22 Test

For each of the following, mark the letter of the best choice.

**1.** Kenya's economy mainly depends on tourism and

○ **A** agriculture.
○ **B** mining.
○ **C** shipping.
○ **D** imperialism.

**2.** Which country was assisted by the United Nations in the 1990s after a civil war and severe drought?

○ **A** Somalia
○ **B** Tanzania
○ **C** Uganda
○ **D** Ethiopia

**3.** Mount Kilimanjaro is covered in snow and ice because of its

○ **A** moist climate.
○ **B** rugged highlands.
○ **C** low latitude.
○ **D** high elevation.

**4.** What is genocide?

○ **A** a policy of colonization
○ **B** restrictions on international shipping
○ **C** the intentional destruction of a people
○ **D** ineffective government policies

**5.** Before it gained independence in 1993, Eritrea was a

○ **A** German colony.
○ **B** vast empire.
○ **C** Muslim state.
○ **D** province of Ethiopia.

**6.** A landform in East Africa formed by the movement of tectonic plates is the

○ **A** Great Rift Valley.
○ **B** Serengeti Plain.
○ **C** Sudan Basin.
○ **D** Nubian Desert.

**7.** Who brought Islam to East Africa?

○ **A** Issa
○ **B** French
○ **C** Arabs
○ **D** Germans

**8.** The belief that the natural world contains spirits is held by followers of

- ○ **A** Islam.
- ○ **B** animist religions.
- ○ **C** Christianity.
- ○ **D** Swahili.

**9.** The Nile makes it possible for the Sudanese to

- ○ **A** protect their borders.
- ○ **B** prevent ethnic conflicts.
- ○ **C** trade with countries along the eastern coast.
- ○ **D** irrigate crops in the desert.

**10.** Why did European countries establish colonies in Africa?

- ○ **A** to search for the source of the Nile
- ○ **B** to spread Christianity throughout the region
- ○ **C** to have access to natural resources such as gold, ivory, and rubber
- ○ **D** to help preserve wildlife on the Serengeti Plain

**11.** What sometimes results from the unpredictable rainfall in East Africa?

- ○ **A** irrigation
- ○ **B** drought
- ○ **C** rain shadows
- ○ **D** rifts

**12.** Which East African country was never colonized?

- ○ **A** Ethiopia
- ○ **B** Djibouti
- ○ **C** Kenya
- ○ **D** Rwanda

Name ______________________ Class ______________ Date ______________

# Central Africa

## Chapter 23 Test

For each of the following, mark the letter of the best choice.

**1.** Where are many countries of Central Africa located?
- ○ **A** near the equator
- ○ **B** along the coast of the Indian Ocean
- ○ **C** along the coast of the Pacific Ocean
- ○ **D** along the Zambezi River

**2.** An important transportation route to the interior of the region is the
- ○ **A** Kinshasa highway.
- ○ **B** Cameroon canal.
- ○ **C** coastal railroad.
- ○ **D** Congo River.

**3.** What is a major barrier to travel on Central Africa's main rivers?
- ○ **A** flooding
- ○ **B** rapids and waterfalls
- ○ **C** mudslides
- ○ **D** drought

**4.** What climate supports the region's tropical forests?
- ○ **A** hot, dry climate
- ○ **B** warm, rainy climate
- ○ **C** distinct dry and wet seasons
- ○ **D** cold, wet climate

**5.** Many Central African countries have been slow to develop their natural resources because of
- ○ **A** poor transportation systems.
- ○ **B** land mines.
- ○ **C** malaria.
- ○ **D** protests by farmers.

**6.** A valuable resource that attracted many Europeans to the region in late 1400s was
- ○ **A** oil.
- ○ **B** copper.
- ○ **C** ivory.
- ○ **D** diamonds.

**7.** When did European countries divide Central Africa into colonies?
- ○ **A** in the 1300s
- ○ **B** in the late 1400s
- ○ **C** in the late 1800s
- ○ **D** after World War II

**8.** What religion is most common in former British colonies?

○ **A** Hinduism
○ **B** Islam
○ **C** Roman Catholicism
○ **D** Protestant Christianity

**9.** Since independence, a problem within many countries in the region has been

○ **A** conflicts with European countries.
○ **B** attacks by people from other regions.
○ **C** fighting among ethnic groups.
○ **D** high taxes and few farmers.

**10.** Which of the following problems has made it hard to build stable governments in the region?

○ **A** lack of foreign aid
○ **B** civil wars
○ **C** lack of natural resources
○ **D** foreign invasions

**11.** Why is the economy of Cameroon growing?

○ **A** It has a stable government.
○ **B** It has ports on the Indian Ocean.
○ **C** It has recently discovered copper resources.
○ **D** It has major coastal highways.

**12.** A key challenge for the region is to stop the spread of

○ **A** communist governments.
○ **B** democratic elections.
○ **C** terrorism.
○ **D** malaria and AIDS.

Name ______________________ Class ______________ Date ______________

## Southern Africa

## Chapter 24 Test

For each of the following, mark the letter of the best choice.

**1.** Which two oceans border Southern Africa?
- ○ **A** Atlantic and Pacific
- ○ **B** Indian and Atlantic
- ○ **C** Indian and Pacific
- ○ **D** Atlantic and Arctic

**2.** The steep face at the edge of a plateau or other raised area is called a(n)
- ○ **A** escarpment.
- ○ **B** pan.
- ○ **C** coastal plain.
- ○ **D** delta.

**3.** The Boers were
- ○ **A** Afrikaner frontier farmers.
- ○ **B** British frontier farmers.
- ○ **C** Zulu frontier farmers.
- ○ **D** Khoisan frontier farmers.

**4.** Two major rivers in Southern Africa are the
- ○ **A** Orange River and the Limpopo River.
- ○ **B** Orange River and the Angola River.
- ○ **C** Limpopo River and the Victoria River.
- ○ **D** Drakensberg River and the Comoros River.

**5.** Which country in Southern Africa has lush vegetation and tropical forests?
- ○ **A** South Africa
- ○ **B** Madagascar
- ○ **C** Mozambique
- ○ **D** Namibia

**6.** The Shona are best known for their stone-walled capital city, called
- ○ **A** Cape Town.
- ○ **B** Great Zimbabwe.
- ○ **C** Johannesburg.
- ○ **D** Windhoek.

**7.** Under apartheid in South Africa, blacks had to live in separate areas known as
- ○ **A** pans.
- ○ **B** velds.
- ○ **C** enclaves.
- ○ **D** townships.

**8.** Southern Africa is rich in mineral resources, including

- ○ **A** gold and copper.
- ○ **B** platinum and diamonds.
- ○ **C** coal and iron ore.
- ○ **D** all of the above.

**9.** The open grassland areas of South Africa are known as

- ○ **A** pans.
- ○ **B** the Namib.
- ○ **C** the veld.
- ○ **D** escarpments.

**10.** The early history of Southern Africa includes the

- ○ **A** Khoisan.
- ○ **B** Bantu.
- ○ **C** Shona.
- ○ **D** all of the above.

**11.** In 1652 the Dutch set up a trade station at a natural harbor near

- ○ **A** the Cape of Good Hope.
- ○ **B** Great Zimbabwe.
- ○ **C** Mozambique.
- ○ **D** Pretoria.

**12.** Penalties imposed by one country on another to force changes in policies are called

- ○ **A** apartheid.
- ○ **B** enclaves.
- ○ **C** isolation.
- ○ **D** sanctions.

## The Indian Subcontinent

## Chapter 25 Test

For each of the following, mark the letter of the best choice.

**1.** In 1971, East Pakistan broke off to become the independent country of
- ○ **A** Nepal.
- ○ **B** Bangladesh.
- ○ **C** Sri Lanka.
- ○ **D** Diwali.

**2.** The first urban civilization on the Indian Subcontinent was the
- ○ **A** Tamil.
- ○ **B** Aryan.
- ○ **C** Mauryan.
- ○ **D** Harappan.

**3.** The worldest highest mountain is
- ○ **A** K2.
- ○ **B** Mount Everest.
- ○ **C** Ghats.
- ○ **D** Hindu Kush.

**4.** Indian troops revolted against British rule in the
- ○ **A** 1500s.
- ○ **B** 1600s.
- ○ **C** 1800s.
- ○ **D** 1900s.

**5.** Indian language and culture was greatly influenced by the
- ○ **A** Sinhalese.
- ○ **B** Tamil.
- ○ **C** Sherpas.
- ○ **D** Aryans.

**6.** More than 70 percent of India's population lives
- ○ **A** on mountain farms.
- ○ **B** in the suburbs.
- ○ **C** in villages.
- ○ **D** in cities.

**7.** What country has a larger population than India?
- ○ **A** the United States
- ○ **B** Russia
- ○ **C** China
- ○ **D** Pakistan

**8.** The Indus River flows through
- ○ **A** India.
- ○ **B** Bangladesh.
- ○ **C** Sri Lanka.
- ○ **D** Pakistan.

**Study the map below and answer the question that follows.**

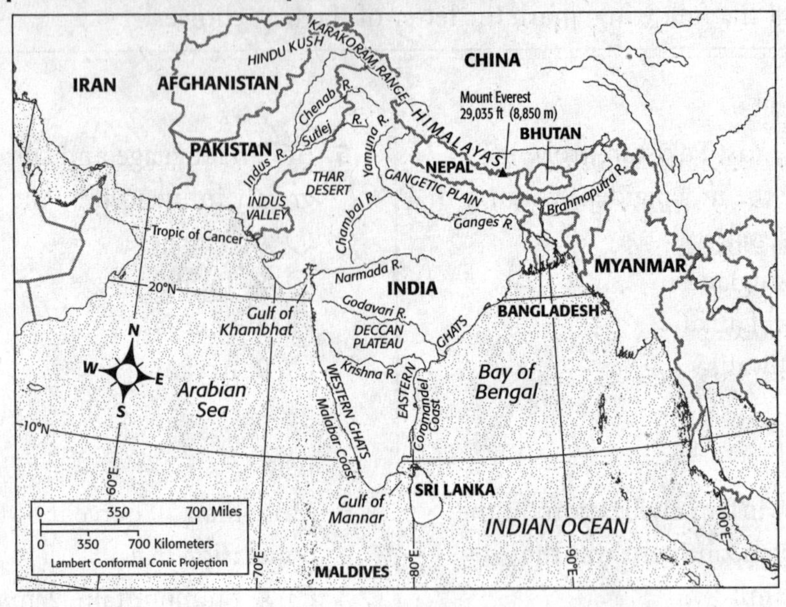


**9.** Which mountain range contains Mt. Everest?

○ **A** Karakoram Range
○ **B** Eastern Ghats
○ **C** Himalayas
○ **D** Hindu Kush

**10.** What system divided Indian society into groups based on birth or occupation?

○ **A** partition
○ **B** caste
○ **C** colonies
○ **D** urbanization

**11.** Which Indian leader helped expand the Mauryan Empire?

○ **A** Asoka
○ **B** Babur
○ **C** Akbar
○ **D** Gandhi

**12.** Under whose rule was the Taj Mahal built?

○ **A** Mughal
○ **B** Aryan
○ **C** Harappan
○ **E** Mauryan

Name ______________________ Class ______________ Date ____________



For each of the following, mark the letter of the best choice.

**1.** Which religion or set of beliefs stresses living simply and in harmony with nature?
- ○ **A** Daoism
- ○ **B** Buddhism
- ○ **C** ancestor worship
- ○ **D** communism

**2.** The southeast is sometimes struck by violent storms called
- ○ **A** monsoons.
- ○ **B** typhoons.
- ○ **C** blizzards.
- ○ **D** sandstorms.

**3.** Most Mongolians make their living as
- ○ **A** horse racers.
- ○ **B** livestock herders.
- ○ **C** rice farmers.
- ○ **D** tourist guides.

**4.** China's first Communist leader was
- ○ **A** Mao Zedong.
- ○ **B** Shi Huangdi.
- ○ **C** Deng Xiaoping.
- ○ **D** Genghis Khan.

**5.** For much of the 20th century, Mongolia was under the influence of
- ○ **A** Japan.
- ○ **B** Taiwan.
- ○ **C** the Soviet Union.
- ○ **D** China.

**6.** China's last dynasty was the
- ○ **A** Qing.
- ○ **B** Shang.
- ○ **C** Beijing.
- ○ **D** Zhou.

**7.** Many countries consider China's response to rebellions as violations of
- ○ **A** trade agreements.
- ○ **B** human rights.
- ○ **C** free enterprise principles.
- ○ **D** environmental protections.

**8.** The most dominant influence on Taiwan's culture today is

○ **A** Chinese.
○ **B** Japanese.
○ **C** Mongolian.
○ **D** European.

**9.** Attempts to modernize China's economy after 1976 were begun by

○ **A** Chiang Kai-shek.
○ **B** Mao Zedong.
○ **C** Deng Xiaoping.
○ **D** the Dalai Lama.

**10.** What led to the events of Tiananmen Square in 1989?

○ **A** People rebelled in Tibet.
○ **B** Foreign businesses had begun to own Chinese companies.
○ **C** Protesters demanded political rights and freedoms.
○ **D** Countries threatened to limit or stop trade with China.

**11.** The Three Gorges Dam will create power from

○ **A** water.
○ **B** wind.
○ **C** gasoline.
○ **D** coal.

**12.** Most Chinese live in the

○ **A** Taklimakan Desert.
○ **B** North China Plain.
○ **C** Plateau of Tibet.
○ **D** Sichuan Basin.

Name ______________________ Class ______________ Date ____________

## Japan and the Koreas

## Chapter 27 Test

For each of the following, mark the letter of the best choice.

**1.** What religion was brought to Korea and later carried to Japan by Chinese missionaries?
- ○ **A** Shinto
- ○ **B** Buddhism
- ○ **C** Confucianism
- ○ **D** Christianity

**2.** Most of Japan's major cities are located on the island of
- ○ **A** Honshu.
- ○ **B** Hokkaido.
- ○ **C** Shikoku.
- ○ **D** Kyushu.

**3.** About how many people live in Pyongyang?
- ○ **A** 300,000
- ○ **B** 1,000,000
- ○ **C** 3,000,000
- ○ **D** 10,000,000

**4.** The average farm in the U.S. is how many times as large as the average farm in Japan?
- ○ **A** 5
- ○ **B** 15
- ○ **C** 50
- ○ **D** 175

**5.** The Korean War began in
- ○ **A** 1941.
- ○ **B** 1950.
- ○ **C** 1963.
- ○ **D** 1994.

**6.** In Japan, the elected legislature is called the
- ○ **A** Kyoto Protocol.
- ○ **B** Diet.
- ○ **C** Parliament.
- ○ **D** Shogun.

**7.** What type of characters represent whole words in the Japanese writing system?
- ○ **A** Kanji
- ○ **B** Kami
- ○ **C** Kana
- ○ **D** Kimchi

**8.** Compared to California, Japan is
- ○ **A** half as large, with twice the population.
- ○ **B** twice as large, with half the population.
- ○ **C** about the same size, with four times the population.
- ○ **D** four times as large, with about the same population.

**9.** The Democratic People's Republic of Korea (North Korea) is really
- ○ **A** a republic.
- ○ **B** a democracy.
- ○ **C** a democracy and a republic.
- ○ **D** neither a democracy nor a republic.

**10.** Kimchi is made from pickled
- ○ **A** cabbage.
- ○ **B** eggs.
- ○ **C** fish.
- ○ **D** beets.

**11.** About what percent of South Koreans are Christian?
- ○ **A** 5 percent
- ○ **B** 10 percent
- ○ **C** 25 percent
- ○ **D** 60 percent

**12.** Japan's successful trade policy has created a
- ○ **A** large food supply.
- ○ **B** strong work ethic.
- ○ **C** huge trade surplus.
- ○ **D** shortage of raw materials.

Name ______________________ Class ______________ Date ____________

## Southeast Asia

## Chapter 28 Test

For each of the following, mark the letter of the best choice.

**1.** Mainland Southeast Asia has mainly a

- ○ **A** tropical savanna climate.
- ○ **B** highland climate.
- ○ **C** monsoon climate.
- ○ **D** steppe climate.

**2.** What is the capital of Myanmar?

- ○ **A** Bangkok
- ○ **B** Phnom Penh
- ○ **C** Hanoi
- ○ **D** Yangon

**3.** The supreme ruler of a Muslim country, such as Brunei, is a

- ○ **A** president.
- ○ **B** sultan.
- ○ **C** monk.
- ○ **D** monarch.

**4.** Which country did the United States grant independence to after World War II?

- ○ **A** Myanmar
- ○ **B** Thailand
- ○ **C** the Philippines
- ○ **D** Singapore

**5.** What are the two peninsulas of Southeast Asia?

- ○ **A** Philippine Peninsula and Thailand Peninsula
- ○ **B** Indochina Peninsula and Malay Peninsula
- ○ **C** Myanmar Peninsula and Malay Peninsula
- ○ **D** Indochina Peninsula and Vietnam Peninsula

**6.** The main religions of Southeast Asia are Buddhism, Christianity, Hinduism, and

- ○ **A** animism.
- ○ **B** Judaism.
- ○ **C** Islam.
- ○ **D** Jainism.

**7.** Although a few Filipinos are wealthy, most are poor

- ○ **A** fishers.
- ○ **B** factory workers.
- ○ **C** shop keepers.
- ○ **D** farmers.

**8.** Many of the plants and animals of Southeast Asia's tropical rain forests are endangered because of

○ **A** biodiversity.
○ **B** deforestation.
○ **C** flooding.
○ **D** tsunamis.

**9.** Some countries will not trade with Myanmar because of its poor record on

○ **A** human rights.
○ **B** environmental issues.
○ **C** religious rights.
○ **D** economic issues.

**10.** After Dutch traders ousted the Portuguese in the 1600s and 1700s, Portugal kept only the small island of

○ **A** Java.
○ **B** Borneo.
○ **C** Timor.
○ **D** Luzon.

**11.** What is the main reason why Singapore is a rich country?

○ **A** It has lots of rich farmland.
○ **B** It is a major tourist center.
○ **C** It is located near the United States.
○ **D** It is located on a major shipping route.

**12.** What causes tsunamis?

○ **A** monsoons and typhoons
○ **B** underwater earthquakes and volcanic eruptions
○ **C** melting glaciers
○ **D** deforestation and flooding

# The Pacific World

## Chapter 29 Test

For each of the following, mark the letter of the best choice.

**1.** What protects Earth's living things from the harmful effects of the sun's ultraviolet rays?
- ○ **A** global warming
- ○ **B** the ozone layer
- ○ **C** Antarctic ice shelves
- ○ **D** the polar desert

**2.** What Pacific Island region includes New Guinea and Fiji?
- ○ **A** Cook Islands
- ○ **B** Melanesia
- ○ **C** Marshall Islands
- ○ **D** Polynesia

**3.** What type of climate does most of Antarctica have?
- ○ **A** highland
- ○ **B** marine
- ○ **C** steppe
- ○ **D** ice cap

**4.** The Southern Alps are a key feature of
- ○ **A** South Island.
- ○ **B** Tasmania.
- ○ **C** North Island.
- ○ **D** Kiribati.

**5.** An atoll is a
- ○ **A** volcanic island.
- ○ **B** coral island.
- ○ **C** sandstone island.
- ○ **D** continental island.

**6.** What did the Antarctic Treaty of 1959 ban?
- ○ **A** mining
- ○ **B** tourism
- ○ **C** scientific research
- ○ **D** military activity

**7.** Which of the following is a typical feature of low islands?
- ○ **A** coconut palms
- ○ **B** rugged mountains
- ○ **C** fertile soil
- ○ **D** dense forests

**8.** Who were the first humans to live in Australia?

○ **A** Maori
○ **B** European settlers
○ **C** Aborigines
○ **D** British prisoners

**9.** Ice shelves in Antarctica form

○ **A** on mountain tops.
○ **B** along the coast.
○ **C** in the interior valleys.
○ **D** on floating icebergs.

**10.** Australia and New Zealand are the top producers of

○ **A** rice.
○ **B** wool.
○ **C** oil.
○ **D** heavy machines.

**11.** Australia's two largest cities are

○ **A** Wellington and Auckland.
○ **B** Auckland and Sydney.
○ **C** Port Moresby and Melbourne.
○ **D** Sydney and Melbourne.

**12.** Papua New Guinea is the eastern part of a

○ **A** high island.
○ **B** coral island.
○ **C** low island.
○ **D** Polynesian island.

# Final Test

# Final Test

For each of the following, mark the letter of the best choice.

**1.** Geographers look at the world
- ○ **A** by studying cities first.
- ○ **B** at the local, regional, and global levels.
- ○ **C** by studying only its physical features.
- ○ **D** as separate regions with no effect on each other.

**2.** A geographer's tools include
- ○ **A** maps and globes.
- ○ **B** satellite images.
- ○ **C** notebooks and tape recorders.
- ○ **D** all of the above.

**3.** The two main branches of geography are
- ○ **A** regional and local.
- ○ **B** cartography and meteorology.
- ○ **C** the study of water and the study of landforms.
- ○ **D** physical geography and human geography.

**4.** Which theme of geography describes features that make a site unique?
- ○ **A** Human-environment interaction
- ○ **B** Movement
- ○ **C** Place
- ○ **D** Location

**5.** Which theory suggests that the continents were once part of one supercontinent?
- ○ **A** tilt and rotation
- ○ **B** continental drift
- ○ **C** Ring of Fire
- ○ **D** eruption patterns

**6.** The main processes of the water cycle are
- ○ **A** evaporation and precipitation.
- ○ **B** drought and flooding.
- ○ **C** salt water and freshwater.
- ○ **D** groundwater and surface water.

**7.** Which of the following creates Earth's change in seasons?

- ○ **A** Earth's tilt
- ○ **B** Earth's rotation
- ○ **C** continental drift
- ○ **D** latitude

**8.** Earth's glaciers can cause

- ○ **A** precipitation.
- ○ **B** drought.
- ○ **C** erosion.
- ○ **D** condensation.

**9.** Which of the following is a renewable resource?

- ○ **A** natural gas
- ○ **B** a gemstone
- ○ **C** a forest
- ○ **D** petroleum

**10.** What happens when an environment changes?

- ○ **A** Habitats are often destroyed.
- ○ **B** Prevailing winds shift to the opposite direction.
- ○ **C** Nonrenewable resources are used even faster.
- ○ **D** Humans fix the problem by introducing new species.

**11.** Ecosystems

- ○ **A** are any materials found in nature that people use or value.
- ○ **B** are made up of renewable and nonrenewable resources.
- ○ **C** can be of any size and can occur whenever air, water, and soil support life.
- ○ **D** move heat around Earth through large streams of surface water.

**12.** Which type of climate can occur at many different latitudes?

- ○ **A** tundra
- ○ **B** highland
- ○ **C** tropical savanna
- ○ **D** subarctic

Name ______________________ Class ______________ Date ______________



**Study the map below and answer the question that follows.**

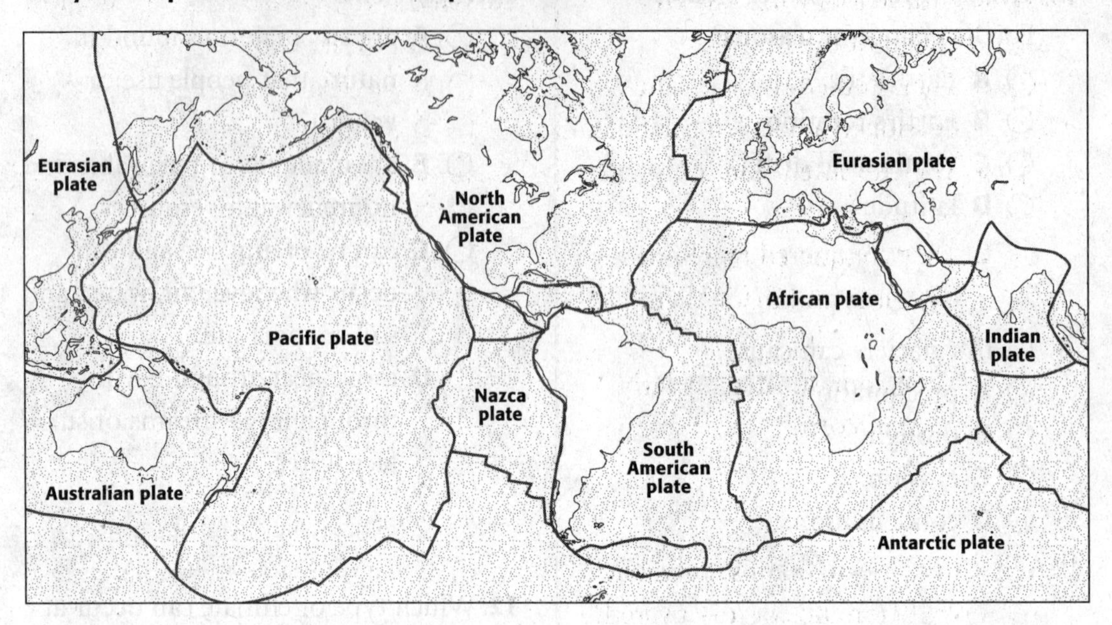


**13.** Two continental plates that seem to have collided are the

- ○ **A** North American plate and Antarctic plate.
- ○ **B** African plate and Australian plate.
- ○ **C** Eurasian plate and Indian plate.
- ○ **D** South American plate and Pacific plate.

**14.** A group of people who share a common culture and ancestry are a(n)

- ○ **A** culture region.
- ○ **B** ethnic group.
- ○ **C** democracy.
- ○ **D** culture trait.

**15.** Which is the economic activity that uses raw materials to produce or manufacture something new?

- ○ **A** primary industry
- ○ **B** secondary industry
- ○ **C** tertiary industry
- ○ **D** quaternary industry

**16.** What is the process of moving from one place to live in another?

- ○ **A** globalization
- ○ **B** cultural diffusion
- ○ **C** migration
- ○ **D** innovation

**17.** Which of the following was an achievement of the Maya?

- ○ **A** They studied the stars and developed a detailed calendar.
- ○ **B** They invented iron tools and carts.
- ○ **C** They conquered neighboring tribes and made them pay taxes.
- ○ **D** They domesticated an early form of corn.

**18.** What were Mexico's most valuable resources before oil was discovered?

- ○ **A** coal and natural gas
- ○ **B** scenic landscapes
- ○ **C** rich soils
- ○ **D** minerals

**19.** The economic development of Central America is limited because it

- ○ **A** lies in an earthquake zone.
- ○ **B** is threatened by hurricanes.
- ○ **C** depends on sugarcane.
- ○ **D** has few energy resources.

**20.** What is the official language in most Central American countries?

- ○ **A** English
- ○ **B** Spanish
- ○ **C** French
- ○ **D** Creole

**21.** The countries in Caribbean South America that export oil are

- ○ **A** Venezuela and Guyana.
- ○ **B** Colombia and Suriname.
- ○ **C** Venezuela and Colombia.
- ○ **D** Colombia and Guyana.

**22.** Venezuela won independence from Spain in the

- ○ **A** 1600s.
- ○ **B** 1700s.
- ○ **C** 1800s.
- ○ **D** 1900s.

**23.** The Amazon Basin is a

- ○ **A** rugged plateau.
- ○ **B** giant floodplain.
- ○ **C** huge estuary.
- ○ **D** dry desert.

**24.** Which country in Atlantic South America is landlocked?

- ○ **A** Brazil
- ○ **B** Argentina
- ○ **C** Uruguay
- ○ **D** Paraguay

**25.** How does the economy of Chile compare with the economies of other Pacific South American countries?

- ○ **A** It has the strongest economy in the region.
- ○ **B** It is the only country to rely on oil exports.
- ○ **C** It has the highest poverty rate in the region.
- ○ **D** It is the only country that does not participate in international trade.

**26.** Which of the following is an effect of El Niño?

- ○ **A** Fish swarm to affected areas.
- ○ **B** The cool water of the Pacific Ocean warms up.
- ○ **C** Rivers cut through dry coastal regions.
- ○ **D** Areas along the coast experience drought.

**27.** What large island lies south of the Straight of Magellan?

- ○ **A** Paraguay
- ○ **B** the altiplano
- ○ **C** La Paz
- ○ **D** Tierra del Fuego

**28.** All but two states border each other and make up the main part of the United States. These two states are

- ○ **A** Maine and Vermont.
- ○ **B** Florida and Georgia.
- ○ **C** Texas and New Mexico.
- ○ **D** Alaska and Hawaii.

**29.** American movies, television programs, and sports are examples of

- ○ **A** diversity.
- ○ **B** popular culture.
- ○ **C** immigration.
- ○ **D** all of the above.

**30.** The building of the transcontinental railroad helped Canada by

- ○ **A** increasing movement to cities.
- ○ **B** attracting more immigrants from France.
- ○ **C** linking British Columbia with eastern provinces.
- ○ **D** giving Native Canadians their own government in Nunavut.

**31.** Which physical feature do the United States and Canada share?

○ **A** Hudson Bay
○ **B** Rocky Mountains
○ **C** Mississippi River
○ **D** Lake Winnipeg

**32.** What are the highest mountains in Europe?

○ **A** the Apennines
○ **B** the Pindus
○ **C** the Pyrenees
○ **D** the Alps

**33.** Which of the following describes the climate in much of southern Europe?

○ **A** warm, dry summers and mild, wet winters
○ **B** warm, wet summers and mild, dry winters
○ **C** warm, wet summers and harsh, wet winters
○ **D** mild, dry summers and harsh, dry winters

**34.** What is the westernmost peninsula in Europe?

○ **A** Iberian
○ **B** Balkan
○ **C** Italy
○ **D** Pindus

**35.** What landform stretches from the Atlantic coast into Eastern Europe?

○ **A** Alps
○ **B** Jura Mountains
○ **C** Northern European Plain
○ **D** Balkan Peninsula

**36.** During World War II, France and the Benelux Countries were invaded by

○ **A** Austria.
○ **B** Germany.
○ **C** Poland.
○ **D** Russia.

**37.** All of the following countries belong to the European Union *except*

○ **A** Luxembourg.
○ **B** France.
○ **C** Switzerland.
○ **D** Germany.

**38.** The physical geography of Northern Europe

○ **A** changes dramatically in different locations.
○ **B** is similar throughout the region.
○ **C** is similar to Southern Europe.
○ **D** is similar to West-Central Europe.

**39.** All of the Scandinavian countries

○ **A** have strong economies and high standards of living.

○ **B** have refused to join the European Union.

○ **C** use geothermal energy for heating homes.

○ **D** lag far behind in the citizens' education.

**40.** Which of the following make up the United Kingdom?

○ **A** England, Wales, Iceland, and Northern Ireland

○ **B** England, Scotland, Wales, and Northern Ireland

○ **C** Iceland, England, Greenland, and Northern Ireland

○ **D** Ireland, Denmark, Greenland, and Wales

**41.** What body of water is most important to trade and transportation throughout Eastern Europe?

○ **A** Rhine

○ **B** Danube

○ **C** Vistula

○ **D** Aegean Sea

**42.** Much of the violence that occurred in the Balkans in the 1990s resulted from

○ **A** Soviet domination.

○ **B** poor economic planning.

○ **C** ethnic conflict.

○ **D** mistreatment of Magyars.

**43.** During the fall of the Soviet Union, its leader was

○ **A** Ivan the Terrible.

○ **B** Vladimir Lenin.

○ **C** Joseph Stalin.

○ **D** Mikhail Gorbachev.

**44.** What two continents meet in Russia's Ural Mountains?

○ **A** Europe and Antarctica

○ **B** Asia and Africa

○ **C** Europe and Africa

○ **D** Europe and Asia

**45.** Siberia is best described as a

○ **A** small region with a tropical climate.

○ **B** vast area with a harsh climate.

○ **C** flat, marshy land with a mild climate.

○ **D** barren plain with a dry, desert climate.

**46.** Much of Syria and Jordan are covered by the
- ○ **A** Syrian Desert.
- ○ **B** Negev.
- ○ **C** Pontic Mountains.
- ○ **D** Taurus Mountains.

**47.** The series of invasions of Palestine launched by Christians from Europe was called
- ○ **A** the Diaspora.
- ○ **B** the Crusades.
- ○ **C** Zionism.
- ○ **D** Knesset.

**48.** Which country shares a border with both the Mediterranean and Black seas?
- ○ **A** Turkey
- ○ **B** Israel
- ○ **C** Lebanon
- ○ **D** Syria

**49.** The country of Jordan was created after
- ○ **A** the fall of the Roman Empire.
- ○ **B** the fall of the Ottoman Empire.
- ○ **C** World War I.
- ○ **D** World War II.

**50.** The Eastern Mediterranean lies between the continents of
- ○ **A** South America and North America.
- ○ **B** Asia and Australia.
- ○ **C** Europe and Asia.
- ○ **D** Africa and Europe.

**51.** When was Palestine declared to be the nation of Israel?
- ○ **A** the late 1500s
- ○ **B** the late 1800s
- ○ **C** 1916
- ○ **D** 1948

**52.** Which of the following is found on the Arabian Peninsula?
- ○ **A** the largest sand desert in the world
- ○ **B** the largest mountain range in the world
- ○ **C** permanent rivers
- ○ **D** tropical forests

**53.** Which country is mostly covered with plateaus and mountains?
- ○ **A** Iraq
- ○ **B** Iran
- ○ **C** Yemen
- ○ **D** Oman

**54.** One influence of Islam on the countries of the Arabian Peninsula is that

- ○ **A** boys and girls go to schools together.
- ○ **B** many women own and run businesses.
- ○ **C** men and women wear clothes that cover their arms and legs.
- ○ **D** women often appear in public alone.

**55.** The world's first civilization was located in present-day

- ○ **A** Iran.
- ○ **B** Iraq.
- ○ **C** Saudi Arabia.
- ○ **D** Oman.

**56.** Which country is made up of a group of islands?

- ○ **A** Kuwait
- ○ **B** the United Arab Emirates
- ○ **C** Qatar
- ○ **D** Bahrain

**57.** Who has ruled Saudi Arabia since 1932?

- ○ **A** ayatollahs
- ○ **B** Persians
- ○ **C** an elected legislature
- ○ **D** members of the Saud family

**58.** Which country shares a large border with Russia?

- ○ **A** Afghanistan
- ○ **B** Turkmenistan
- ○ **C** Uzbekistan
- ○ **D** Kazakhstan

**59.** Which statement best explains why people speak many different languages in Central Asia?

- ○ **A** Each government established an official language.
- ○ **B** Each religion requires its people to speak a certain language.
- ○ **C** Each ethnic group speaks its own language.
- ○ **D** The Soviets encouraged people to continue using their traditional languages.

**60.** What is a yurt?

- ○ **A** a large apartment building in a city
- ○ **B** a moveable home used by nomads
- ○ **C** a traditional religious building
- ○ **D** a small mountain village

**61.** Which word best describes Central Asia's physical geography?

- ○ **A** coastal
- ○ **B** fertile
- ○ **C** landlocked
- ○ **D** tropical

**62.** What type of climate does most of North Africa have?

- ○ **A** tropical
- ○ **B** Mediterranean
- ○ **C** steppe
- ○ **D** desert

**63.** What is the Maghreb?

- ○ **A** the Atlas Mountain region of the Sahara
- ○ **B** a narrow area along the western coast of North Africa
- ○ **C** western Libya, Tunisia, Algeria, and Morocco
- ○ **D** a series of large oases in central North Africa

**64.** When did Arab powers from Southwest Asia rule North Africa?

- ○ **A** during the 2500s BC
- ○ **B** from the AD 600s to the 1800s
- ○ **C** during the mid-1900s
- ○ **D** from the 1800s to today

**65.** What religion do most North Africans practice?

- ○ **A** Islam
- ○ **B** Christianity
- ○ **C** Judaism
- ○ **D** Buddhism

**66.** The great kingdoms of West Africa became powerful by

- ○ **A** conquering Mediterranean countries.
- ○ **B** establishing colonies in northern Africa.
- ○ **C** using ships to send goods around the world.
- ○ **D** controlling trade routes across the Sahara.

**67.** The Atlantic slave trade

- ○ **A** supplied labor for European colonies.
- ○ **B** supplied slaves to North Africa.
- ○ **C** was opposed by most European countries.
- ○ **D** resulted in many Africans being shipped to Europe.

**68.** Political boundaries drawn by European colonizers

- ○ **A** generally followed climate zones.
- ○ **B** kept people who spoke the same language together.
- ○ **C** kept people of the same religion together.
- ○ **D** often separated members of the same ethnic group.

**69.** Kenya's economy depends mainly on tourism and

- ○ **A** agriculture.
- ○ **B** mining.
- ○ **C** shipping.
- ○ **D** imperialism.

**70.** The Nile makes it possible for the Sudanese to

- ○ **A** protect their borders.
- ○ **B** prevent ethnic conflicts.
- ○ **C** irrigate crops in the desert.
- ○ **D** trade with countries along the eastern coast.

**71.** When did European powers divide Central Africa into colonies?

- ○ **A** in the 1300s.
- ○ **B** in the late 1400s.
- ○ **C** in the late 1800s.
- ○ **D** after World War II.

**72.** Tourists come to Tanzania to

- ○ **A** mine gold and diamonds.
- ○ **B** go on safaris.
- ○ **C** buy tea and coffee.
- ○ **D** explore the Nile.

**73.** Which of the following is a major barrier to travel on Central Africa's main rivers?

- ○ **A** flooding
- ○ **B** rapids and waterfalls
- ○ **C** mudslides
- ○ **D** drought

**74.** Since independence, a problem within many former African colonies has been

- ○ **A** conflicts with European countries.
- ○ **B** attacks by people from other regions.
- ○ **C** fighting among ethnic groups.
- ○ **D** high taxes and few farmers.

**75.** The steep face at the edge of a plateau or other raised area is called a(n)

- ○ **A** delta.
- ○ **B** pan.
- ○ **C** coastal plain.
- ○ **D** escarpment.

**76.** The Boers were

- ○ **A** Afrikaner frontier farmers.
- ○ **B** British frontier farmers.
- ○ **C** Zulu frontier farmers.
- ○ **D** Khoisan frontier farmers.

**77.** What was South Africa's past policy of separation of races called?

- ○ **A** townships
- ○ **B** enclaves
- ○ **C** sanctions
- ○ **D** apartheid

**78.** What is the dominant religion of India today?

- ○ **A** Islam
- ○ **B** Christianity
- ○ **C** Hinduism
- ○ **D** Buddhism

**79.** What did the British agree to in 1947?

- ○ **A** independence for Bangladesh
- ○ **B** freeing Ghandi from prison
- ○ **C** ending the caste system
- ○ **D** the partition of India

**80.** The first urban civilization on the Indian Subcontinent was the

- ○ **A** Harappan.
- ○ **B** Tamil.
- ○ **C** Mauryan.
- ○ **D** Aryan.

**81.** Taiwan's climate most resembles that of

- ○ **A** northern China.
- ○ **B** western Mongolia.
- ○ **C** central Tibet.
- ○ **D** southeastern China.

**82.** For much of the 20th century Mongolia was under the influence of

- ○ **A** Japan.
- ○ **B** Taiwan.
- ○ **C** the Soviet Union.
- ○ **D** China.

**83.** Most Chinese live in the

- ○ **A** Taklimakan Desert.
- ○ **B** North China Plain.
- ○ **C** Plateau of Tibet.
- ○ **D** Sichuan Basin.

**84.** After World War II, North Korea formed a government with the help of

- ○ **A** China.
- ○ **B** the United Nations.
- ○ **C** the United States.
- ○ **D** the Soviet Union.

**85.** Unlike Japan, Korea has

- ○ **A** some large plains.
- ○ **B** rugged mountains.
- ○ **C** summer typhoons.
- ○ **D** destructive earthquakes.

**86.** What religion or belief system was brought to Korea and later carried to Japan by Chinese missionaries?

- ○ **A** Shinto
- ○ **B** Buddhism
- ○ **C** Confucianism
- ○ **D** Christianity

**87.** Compared to California, Japan is

- ○ **A** half as large, with twice the population.
- ○ **B** twice as large, with half the population.
- ○ **C** about the same size, with four times the population.
- ○ **D** four times as large, with about the same population.

**88.** What are the two peninsulas of Southeast Asia?

- ○ **A** Philippine Peninsula and Thailand Peninsula.
- ○ **B** Indochina Peninsula and Malay Peninsula.
- ○ **C** Myanmar Peninsula and Malay Peninsula.
- ○ **D** Indochina Peninsula and Vietnam Peninsula.

**89.** What causes tsunamis?

- ○ **A** monsoons and typhoons
- ○ **B** underwater earthquakes and volcanic eruptions
- ○ **C** melting glaciers
- ○ **D** deforestation and flooding

**90.** The main religions of Southeast Asia are Buddhism, Christianity, Hinduism, and

- ○ **A** animism.
- ○ **B** Judaism.
- ○ **C** Islam.
- ○ **D** Jainism.

**91.** What type of climate does most of Antarctica have?

- ○ **A** ice cap
- ○ **B** marine
- ○ **C** steppe
- ○ **D** highland

**92.** Who were the first humans to live in Australia?

- ○ **A** Aborigines
- ○ **B** Maori
- ○ **C** Dutch settlers
- ○ **D** British prisoners

**93.** Which Pacific Island region includes about 2,000 small islands?

- ○ **A** Polynesia
- ○ **B** Melanesia
- ○ **C** Micronesia
- ○ **D** Marshall Islands

Name ______________________ Class ______________ Date ______________

## Final Test

**Study the graph below and answer the question that follows.**

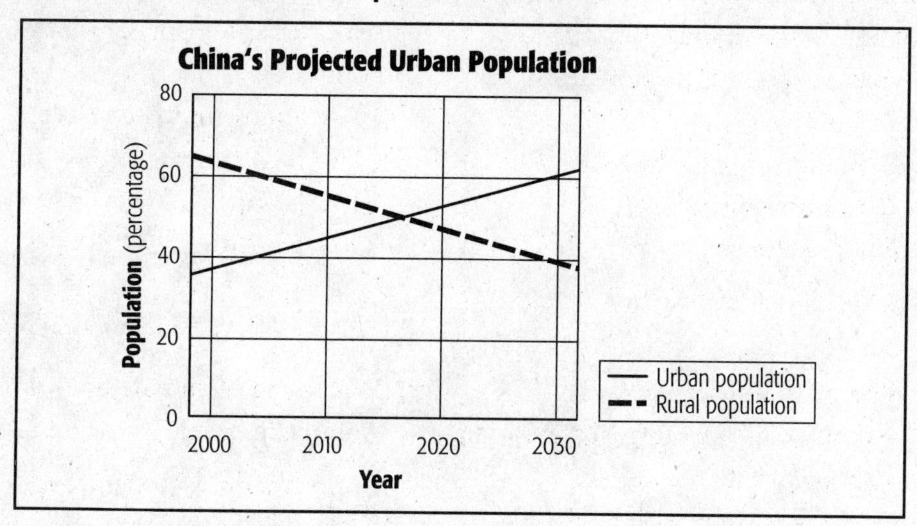


**94.** According to the graph, in about what year will the percentage of China's urban and rural population be the same?

- ○ **A** 2009
- ○ **B** 2017
- ○ **C** 2023
- ○ **D** 2030

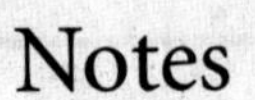

# Notes

# Notes

# Notes